Understanding & Using Radio Communications Receivers

By John Schultz

TAB BOOKS
Blue Ridge Summit, Pa. 17214

FIRST EDITION

FIRST PRINTING—JULY 1972

Printed in the United States
of America

Hardbound Edition: International Standard Book No. 0-8306-2614-X

Paperbound Edition: International Standard Book No. 0-8306-1614-4

Library of Congress Card Number: 72-84080

Cover photo courtesy of The Hallicrafters Co.

(A Subsidiary of NORTHROP CORP.)

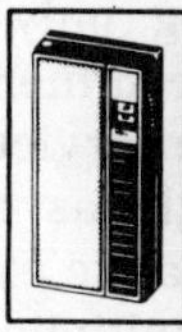

Preface

This handbook is designed to serve the needs of the hobbyist or technician who is interested in shortwave receiving equipment covering the VLF to VHF range, including CB, public safety, and other frequency bands of interest.

The basic circuitry and signal processing functions of a wide variety of receivers are covered in detail. Many examples are given of current receiver design. But, the handbook is not meant to be a "dry" circuitry manual. Chapters are included on understanding propagation as it relates to better reception, receiving antennas, selecting and setting up equipment, examples of professional receiving and monitoring installations, how to interpret the various types of signals heard, where to obtain the frequency and time schedules for all types of stations, etc. In other words, this handbook is intended for the person who actually wants to use and to get the most out of receiving equipment regardless of his degree of technical knowledge.

Radio receiving equipment, whether it be separately installed or a part of two-way radio equipment, and whether it be extremely costly or moderately priced, is often not utilized to its fullest capability. This is probably most true in communications installations using transmitting equipment since attention is primarily devoted to transmitter power level, antenna gain, etc. But, reliable and extended-range two-way radio links are very much dependent upon receiver performance, as well as transmitter performance. There are also several valid reasons why investment in improved receiver performance can produce better communications results more economically than an investment in additional transmitting equipment. For instance, one can modify and experiment with receiving equipment much more freely than with transmitting equipment, where, unless one has the necessary commercial or amateur radio license, adjustments of many sorts cannot legally be made. This is particularly true for the CB operator.

For installations using only receiving equipment, on shortwave frequencies for commercial or hobby use, or on the VHF frequencies for public service uses, the receiver itself is often regarded too strongly as being the total solution to reception problems. The classic example is the investment in a $1,000 receiver with a multiplicity of controls used with a wire antenna costing a few cents. A balanced installation is needed which will prove more economical and render better service. The operator should also understand the electromagnetic environment in which the installation functions. The result will be increased operating pleasure as well as increased station performance.

John J. Schultz

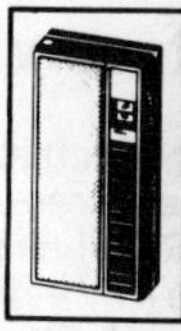

Contents

Some Basics on Radio Wave Propagation

Chapter 1

A detailed knowledge of the mechanics of radio wave propagation is not necessary in using a communications receiver, but a general knowledge of it should be of great help in guiding you to the best of various reception possibilities. In day-to-day reception, the operator is primarily interested in whether or not signals can actually be received from a certain area—propagation forecasting, in other words.

The basics of propagation and propagation forecasting are both covered in this chapter, in addition to some explanation of how the radio spectrum is divided into specific bands for various services. The material in this chapter includes guidance on reception conditions from the basic VLF (Very Low Frequency) region to the VHF (Very High Frequency) region. If these terms are not familiar, please refer briefly to Chapter 2 carefully before reading further.

THE VAST ELECTROMAGNETIC SPECTRUM

The so-called electromagnetic spectrum is a vast region which extends basically from VLF (Very Low Frequencies) to SHF (Super High Frequencies). Each frequency range has its own particular characteristics which determine whether radio waves travel primarily in a direct line from transmitter to receiver, whether they travel over an indirect path or whether they normally travel over short or long distances. VLF and SHF are the extreme designations of the electromagnetic range. Medium wave, short wave, etc., all fall in between. Fig. 1-1 shows how the spectrum is divided. Notice that some portions of the range are designated by a modulation type. The AM (amplitude modulation) broadcast band may be considered medium wave and the FM (frequency modulation) broadcast band is correctly termed a Very High Frequency band. In fact, these designations are normally used for comparable broadcast bands even outside the U.S. Other

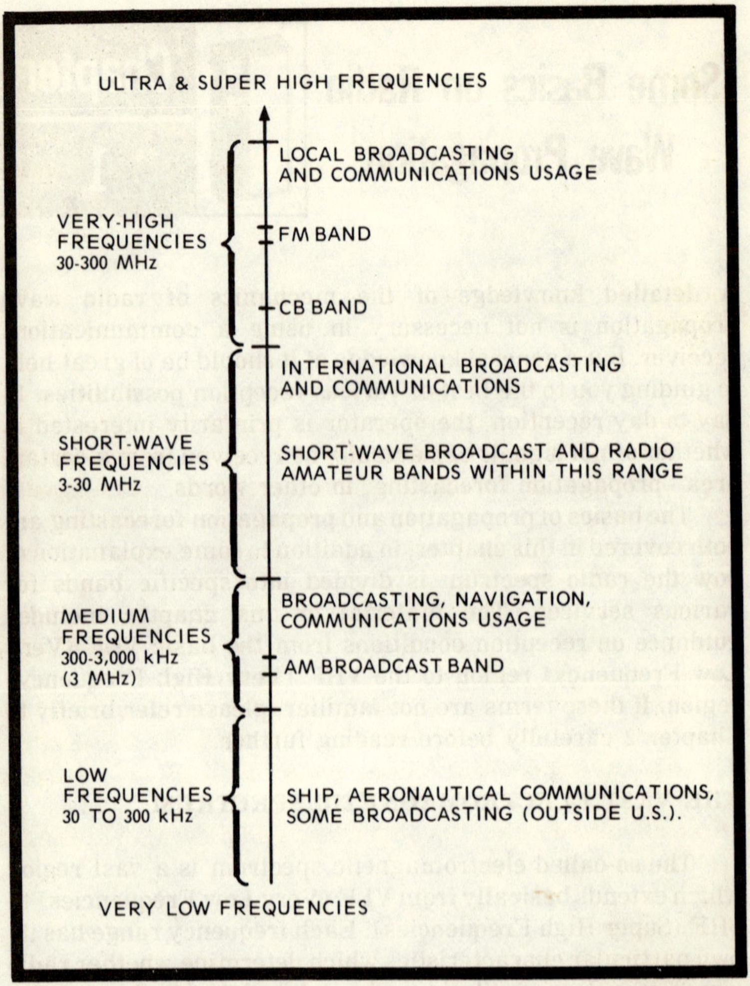

Fig. 1-1. Divisions according to usage in the electromagnetic frequency spectrum.

bands such as the CB (Citizen's Band) are named after the designated use. The main idea to grasp from Fig. 1-1 is that the electromagnetic spectrum is a continuous thing. The AM band does not suddenly stop and the short-wave bands start, for example. However, each band within the spectrum is specifically designated, generally, on the basis the properties that affect propagation. A particular point within the spectrum is designated by its frequency or perhaps wavelength in meters. These terms are explained in the next chapter.

CHARACTERISTICS OF RADIO WAVES

Generation

When an antenna is excited with an alternating current at radio frequencies, rapidly changing electric and magnetic fields are set up around the antenna. This field travels away from the antenna at a velocity (in free space) of 300,000,000 meters per second. It can be thought of as continuously changing back and forth from electric to magnetic fields. Unlike a sound wave which depends upon the movement of air, radio waves can even travel in a vacuum.

The rate at which the fields change is determined by the frequency of the current feeding the antenna. The actual physical distance the wave must travel to go through one complete cycle of change is called wavelength. In free space, one wavelength (λ) in meters is equal to 300 divided by the frequency in MHz. Radio waves of all frequencies are basically the same. The fact that waves of various frequencies can be used for certain methods of communication or over certain distances is only a function of how these waves are affected by the medium in which they travel.

The term "Hertz" (kiloHertz, megaHertz) is used in this text instead of "cycles per second" to express the frequency of a radio wave, since engineering groups have recently adopted this term to honor Heinrich Hertz who in 1886 produced the first man-made electromagnetic or radio waves utilizing a spark-gap apparatus.

Polarization

For convenience, a wave is said to be "polarized" according to the orientation of its electric field to the earth's surface. Also for convenience, the strength of a wave is taken as being the potential length (usually 1 meter) of its electric field. In free space, the strength is inversely proportional to the distance from the transmitting antenna. If distance is doubled, the strength decreases to one fourth. In practice, carefully calibrated receivers and antennas can be used to measure this field strength, which is usually expressed in terms of microvolts per meter. Different communications applications require different field strength levels. A level of 1,000 microvolts per meter may be necessary for broadcast quality, while in a quiet location and with good equipment, a 0.5-microvolt per meter signal may suffice for voice communications.

However, in practice the field strength itself does not determine whether a station can be received in any given location. The determining factor is the relation of the signal strength to the noise present, both natural and man-made, at the receiving location. How much the signal must be above the noise level depends upon the type of communication (on-off keying, AM, FM, etc.) and the quality of the receiving equipment.

Like light waves, another form of electromagnetic radiation even higher in frequency than SHF, radio waves are reflected, refracted and diffracted. Reflection may occur when the reflecting surface is of the right type (bright metal, for instance) and larger than the wavelength of the radio wave. Some "scattering" may occur upon reflection since no surface is perfectly smooth. Any irregularities in the surface will tend to scatter some of the wave.

During refraction, the wave penetrates another medium, but in doing so, its direction of travel is changed. If it encounters, successively, a number of mediums of increasing density, it may be completely turned around.

When a wave encounters an obstacle which it cannot penetrate and which does not reflect it, it may diffract or bend around the obstruction. Generally, low-frequency radio waves will diffract around obstructions more readily than will high-frequency waves; the high-frequency waves reflect off objects much like a light beam.

PROPAGATION PATHS

In the actual physical environment in, on and above the surface of the earth, there are a number of paths by which a radio wave may travel between two points. Usually, one path will predominate, but this does not prevent some energy from traveling in other paths. Antennas can and usually are designed to radiate in a specific propagation path, but the antenna design does not in itself determine the path. These statements may seem obvious, but they are important in understanding how to take advantage of propagation conditions for reception.

Propagation within the earth differs markedly from propagation at the surface and above the surface. Propagation within the earth has generally proved very inefficient and about the only practical application to evolve is some form of buried antenna, designed for bomb-proof or hidden installations, which use surface propagation paths for operation.

GROUND AND DIRECT-WAVE PROPAGATION

Any transmitting antenna will produce a surface wave if excited by an RF current from a transmitter. The strength of this wave diminishes more rapidly than a wave in free space because of absorption by the earth. How rapidly it diminishes depends upon its polarization, frequency and the nature of the terrain over which it travels. Horizontally polarized waves diminish much more rapidly than vertically polarized ones. Most antennas designed to take advantage of this propagation mode are vertically polarized. Higher frequencies are absorbed much more rapidly than lower frequency waves. Also, much less energy is absorbed from the wave if it travels over water rather than over poor soil and irregular terrain. Below about 1,000 kHz, ground-wave coverage can be very extensive, especially with high-powered transmitters.

Within the strong signal area of a station, the surface wave is always present and no fading effects are noticed (the primary service area of an AM broadcast station, for instance). In this area, an antenna that responds to vertically polarized waves is best, although polarization is rarely of importance because of changing surface conditions, as produced by weather and irregular diffraction. Diffraction also causes the wave to "tilt" in the direction of propagation so that an antenna which responds to horizontally tilted waves will prove more effective at extreme distances.

Direct Wave

As the frequency of a radiated wave increases above 30 MHz, absorption of the ground wave increases rapidly and limits communication within a line of slight or direct point-to-point wave. Fig. 1-2 shows the basic forms of ground and direct waves. Direct-wave paths can also exist below 30 MHz, but this would normally be only for situations such as air-to-air or air-to-ground communctions within line-of-sight distances.

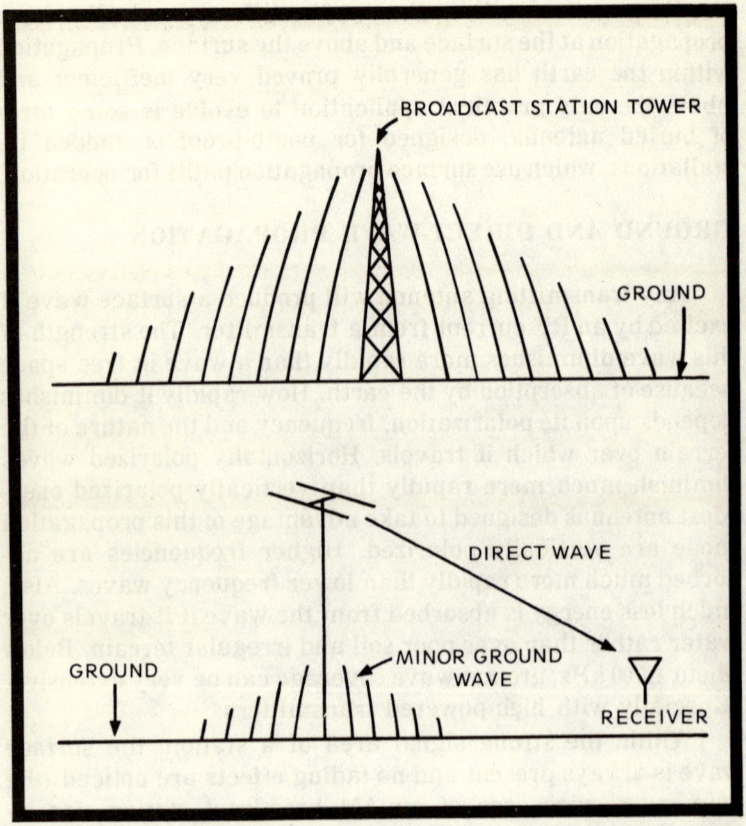

Fig. 1-2. At low frequencies, ground waves alone reach local receivers. At VHF, direct waves reach local receivers.

Depending upon antenna design and siting, there may also exist a ground reflected direct wave. If so, the received signal will be a combination of the direct and reflected signals and range from much stronger to much weaker than the direct wave alone, depending upon whether the signals arrive so as to reinforce or to subtract from each other.

The "radio" line of sight is somewhat more than the straight-line distance because of diffraction of the direct wave, as shown in Fig. 1-3. One interesting characteristic of such transmission is that raising the height of either the receiving or transmitting antenna will more significantly affect the signal strength at any given distance than will an increase in transmitting power. Fig. 1-4 shows a particularly

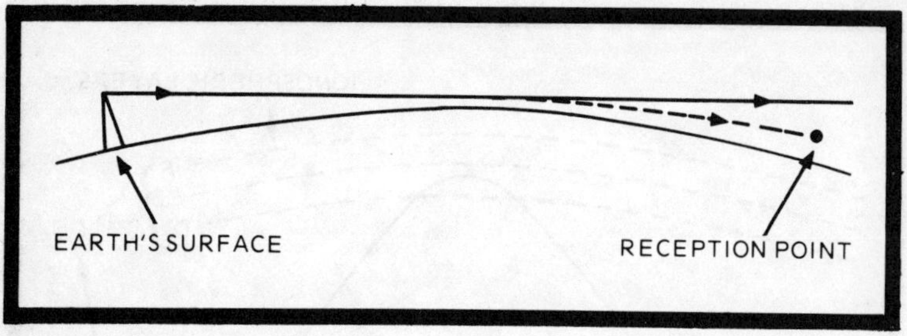

Fig. 1-3. Direct waves bend to allow some reception at extended distances on VHF.

useful graph which can be used to estimate the direct-wave communication distance from a base station to a mobile station. For best results, either by ground or direct wave, the

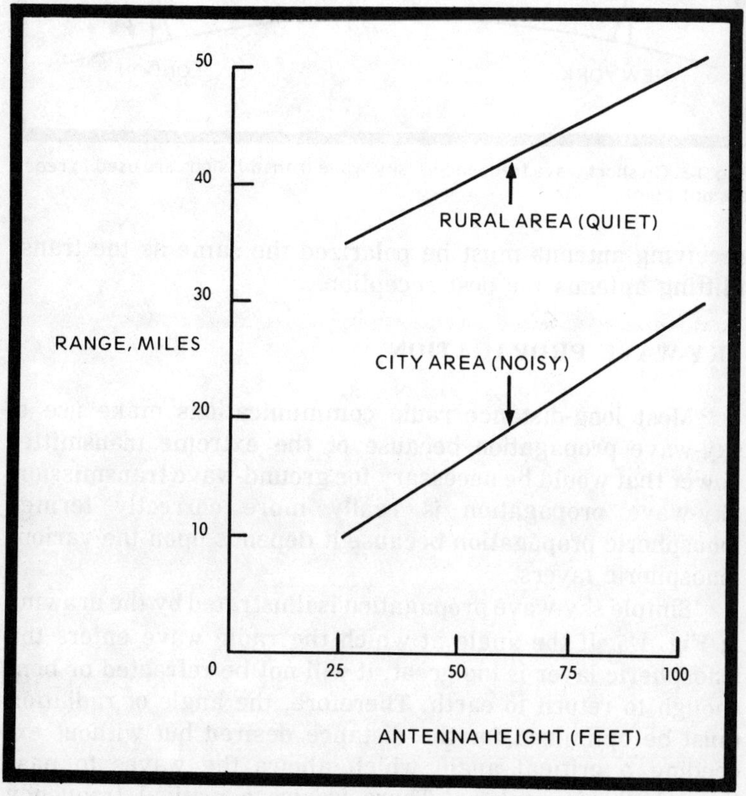

Fig. 1-4. Graph showing maximum distances for base-to-mobile communications at 30 MHz with 50 watts output (transmitter output times antenna gain).

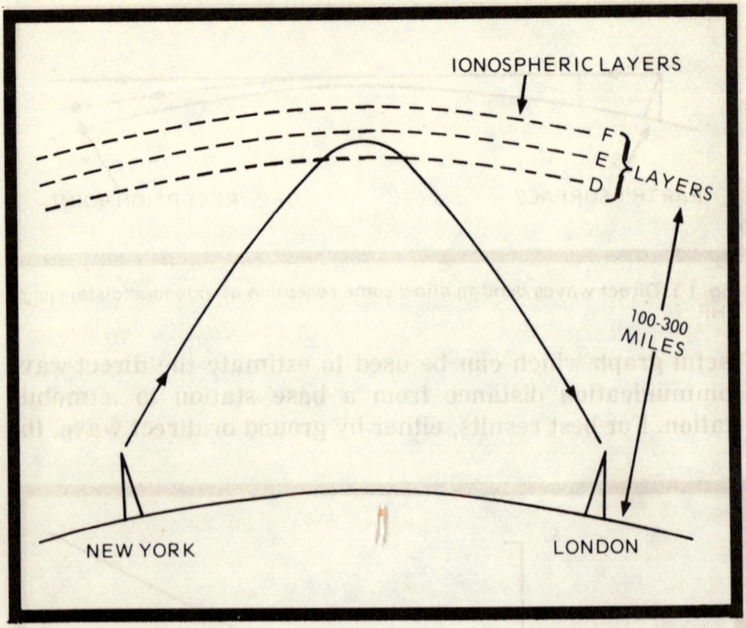

Fig. 1-5. On short-wave frequencies, sky-wave transmissions are used to reach distant points.

receiving antenna must be polarized the same as the transmitting antenna for best reception.

SKY-WAVE PROPAGATION

Most long-distance radio communications make use of sky-wave propagation because of the extreme transmitter power that would be necessary for ground-wave transmission. Sky-wave propagation is really more correctly termed ionospheric propagation because it depends upon the various ionospheric layers.

Simple sky-wave propagation is illustrated by the drawing in Fig. 1-5. If the angle at which the radio wave enters the ionospheric layer is too great, it will not be refracted or bent enough to return to earth. Therefore, the angle of radiation must be optimized for the distance desired but without exceeding a critical angle which allows the waves to pass through the ionosphere. There is also a critical frequency above which the waves penetrate the ionosphere regardless of the angle at which they enter.

A commonly used term is "maximum usable frequency," which is the highest frequency that any given set of ionospheric conditions will refract for a given distance. Lowest usable frequency is the lowest frequency which, considering the variable absorption loss in the ionosphere, will still provide the minimum usable field intensity at a distant point. If a choice of frequencies for reception exists, the best as far as signal strength is concerned will generally be just below the maximum usable frequency, since absorption is at a minimum. Sometimes, such a frequency is referred to as the "optimum working frequency," taken as roughly 20 percent less than the maximum usable frequency. What these frequencies would be at any given time of the day and for any given path, such as from the U.S. to Europe, can be found in many reference books as discussed later in the text.

Generally speaking, sky-wave reception is possible only in those areas where the wave is reflected back to earth. However, some scattering of the signal also occurs upon reflection, and some reception may be possible in parts of the "skip zone" (see Fig. 1-5).

It may appear from Fig. 1-4 that there is only one ionospheric layer. In fact, there are a number of such layers (D, E and F). These layers are composed of ionized gas in the upper atmosphere and strongly influenced by radiation from the sun. The most regular change that occurs is the separation and combination of layers during day and night. One point to remember is that the condition of the ionospheric layer at the refraction point, not at the receiving or transmitting area, determines the effect upon a signal.

The D layer is mainly responsible for certain propagation effects in the 3 to 300-kHz band. The very narrow VLF band (3 to 30 kHz) exhibits a special mode of propagation which includes reflection from the D layer. It is the only portion of the frequency spectrum where a sufficiently high-powered transmitter (such as the 1 million watt Navy stations in Washington and Maine) can put a signal into all portions of the world regardless of ionospheric conditions. There is some reflection of LF signals (30 to 300 kHz). However, at night time this absorption is much less and some reflection occurs from the E layer even though this layer is much weaker at night than during the day. Such reflection can extend the range of an AM broadcast station over a considerable distance so that transoceanic reception of such stations is possible under extremely favorable conditions.

During the daylight hours the E layer is mainly responsible for medium-distance signal paths, although multiple reflections can extend the range by several thousand miles for frequencies up to 10 MHz or more. However, the F layers are the main ones responsible for long-range signal paths in the short-wave range (3 to 30 MHz).

You may wonder about the origin of the letter designations for the ionospheric layers. Sir Edward Appleton, a noted researcher of ionospheric physics, used the letter E for the electric field when studying reflections from the then-called Kennelly-Heavyside layer. He then discovered other layers above and below this layer and used the letters D and F to distinguish them. The letter designations simply became common and remained in use; the layers could just as well be designated 1, 2 and 3.

PROPAGATION CYCLES

At first, an operator may confuse some of the regular cyclic propagation changes with signal disturbances, especially during reception in the short-wave bands. However, the two classes are very distinct. Also, one may sometimes be inclined to call various disturbances a mode of propagation. Although a few disturbances can actually be used to advantage to improve reception temporarily, their occurrence cannot be reasonably predicted.

Short and Long-Term Cycles

Propagation cycles are really sun radiation cycles, since the ionosphere is formed by such radiation. The sun is not a constant source of radiation and, therefore, there are small, continual changes in the ionosphere. These continual changes are partly responsible for short-term fading effects. Any radio wave entering the ionosphere is really a broad front of waves spreading out much like visible light. In a changing ionosphere, these waves are not all acted upon in the same way and the combined effect is a signal that is no longer constant after refraction. Fading effects can be broad in a frequency range or very selective, with only a few hundred cycles difference between signals which fade at different rates.

Long-term fading effects occur as the result of some overall change in the ionosphere. One example is the change between the formation of the night-time and day-time layers; in each case the maximum usable frequency is different, and each varies in density according to the vertical angle of the sun. This daily or diurnal change can be used to interesting advantage at times for both transmitting and reception purposes.

There are three longer term changes in the ionosphere which a receiving operator should be aware of, although some are, at times, so gradual that they may not be recognized immediately. The first is a 27-day cycle associated with the time required for the sun to complete one revolution about its axis. Apparently some particle radiation occurs (which is deflected by the earth's magnetic fields); the term "magnetic storm" is used to describe this occurrence. During maximum sunspot conditions, such storms can be quite severe and last for several days. During such a storm the F layer ion density is reduced, thus reducing the maximum usable frequency. Ion density, particularly in the D layer, and in the E layer, increases, causing increased signal absorption. The maximum usable frequency over a path of several hundred miles in the middle latitudes might fall to less than half. Signal absorption, even on frequencies where F-layer refraction is still possible, might increase tremendously. The effect of all this is that potentially excellent short-wave reception on the higher frequency band may become totally lost.

The second long-term cycle is a seasonal one. During the summer, the F layer is higher than in winter, but has a lower peak ion density. The D, E and F layers have lower peak ion densities in winter. The effect, as far as the maximum usable frequency (muf) is concerned, depends upon the layer being used for communication and the distance covered.

The third and longest term effect noticed caused by the ionosphere is the sunspot cycle which extends over 11 years. Sunspots are actual spots or dark areas of reduced temperature which appear on the surface of the sun. They have been recorded for over a hundred years, using the so-called Wolf sunspot count, and yearly since 1900, including probable values for the next few years. The most pronounced effect of high sunspot counts is the increase in density and height of the F layers. During maximum sunspot periods, low-power, world-wide communication is possible on frequencies as high as 30 MHz during most of the day.

PROPAGATION DISTURBANCES AND UNUSUAL RECEPTION

Propagation disturbances can either drastically limit the reception of signals or cause propagation paths to exist (anomalous propagation) which are not normally there. Disturbances can be of short duration (minutes) or last several days. Some of those which have caused anomalous propagation have lasted even longer. An FM station, for instance, might be received quite clearly over a distance of a hundred miles or more for several days and suddenly, in the space of a few minutes, fade and never be heard again. The main types of disturbances which exist are:

A. Sudden Ionospheric Disturbance (SID) is an effect produced by the eruption in the reddish, incandescent gases around the sun (visible during an eclipse). It is not of the same nature as the 27-day cycle, but its effect upon reception in the short-wave range is the same—greatly increased absorption with a sharp reduction in the maximum usable frequencies. It occurs only in the daytime and may last from several minutes to several hours. The effect is often referred to as the Dellinger Effect after Dr. John Dellinger who discovered the effect in 1935. Usually, a signal affected by such a disturbance will fade out very suddenly and then gradually reappear only as the disturbance ends.

B. Ionospheric storms occur at irregular intervals. They frequently follow from 18 to 36 hours after a sudden ionospheric disturbance and can vary from an insignificant occurrence to one where such tremendous D-layer absorption takes place that no signals whatever are refracted.

C. Sporadic E is the disturbance that, unlike a sudden ionospheric disturbance, may either improve or degrade reception. The cause of sporadic E is not fully understood, but the effect is areas of abnormal ionization in the E layer during the day or night. The maximum usable frequency for long-distance reception may increase to 50 or 60 MHz. This is the effect, for instance, which may allow signals from other countries or from across the U.S. to suddenly be heard on the Citizen Band channels.

D. Meteor trails also cause the propagation of signals beyond normal distances. Meteors, as they enter the atmosphere, produce an ionized region which may refract signals in the 20 to 120-MHz range. The effect upon reception may be much the same as Sporadic E, but the signals which are received have a "flutter" characteristic.

E. "Ducting" is an effect caused by differing temperature layers in the atmosphere or between the earth and the atmosphere. In some areas, such ducts will appear in a periodic manner, but generally their formation is erratic. As the name implies, a duct provides a directional signal path between two points which normally would have no propagation path between them. The duct may be relatively small, and small changes in the receiving antenna location may severely affect reception. Usually, ducting is noted in sea coast areas and not below 50 MHz. Such ducts may extend several hundred miles and the signals received will be relatively steady with a slow, rolling fade. Again, this is an effect that sometimes produces unusually distant communication possibilities on the CB or lower VHF public safety radio channels.

F. Spread F is a disturbance which partially gathers its name from the fact that it spreads out or elongates a signal. For example, if just a pulse of definite length is refracted by an F layer, the returned pulse may be several times longer than the original one. It is caused by irregularities in the ionosphere which scatter a signal, although the exact cause is not clear. Since Spread F is frequently caused by ionospheric storms which occur in the polar regions, the effect upon high-frequency signals is commonly called "auroral flutter." If Spread F is present, it will occur in middle latitudes most frequently around or after midnight. The "flutter" can be so severe at times to prevent the reception of signals in the U.S. from Europe, for instance.

G. Noise is one of the most important disturbances that constantly affect high-frequency communications. There are many varieties of noise, but here we consider only natural noises or static. Man-made noises and noise generated within a receiver are taken up in subsequent chapters.

Natural noises can be classified as follows:

1. Random electrical discharges in the earth's atmosphere produce radio waves over a wide frequency range extending from VLF through UHF. However, the intensity decreases inversely with wavelength until at UHF it is of no significance. Static or atmospherics are the common names for this disturbance.

Electrical discharges in the atmosphere occur all over the earth, but are most intense in various tropical areas. Being like radio waves, they propagate according to ionospheric conditions during the day. Therefore, at any given location, you can receive local as well as distant static. In the middle latitudes, distant static may be more intense than local static. This factor is used to advantage in the static-cancelling antenna described in Chapter 6. Static is generally most intense at night, but it decreases in the daytime because of daytime absorption in the D layer.

2. Cosmic noise is radiation from certain areas in outer space. Only a certain band of frequencies associated with this noise penetrates the ionosphere so that this noise is most predominant in the upper high-frequency to lower VHF range. Unlike static with its clashes and varying intensity, cosmic noise takes the form of a steady hiss as heard in a receiver. The intensity and direction of arrival of this noise varies with time.

3. Precipitation noise or static, usually, is a problem only with moving vehicles, but may also occur at fixed locations during snow, rain or dust storms when high winds are present. The particles striking a surface are ionized by friction, thus the surface accumulates potential and an electrical discharge can occur, producing static. The surface need not be an antenna, but if it is, the static produced may render reception impossible under extreme conditions. Grounding the offending surface or antenna (for direct current) will usually eliminate such static except for large antenna systems where one part of the antenna may develop a charge opposite to that on another section.

PROPAGATION CONDITION FORECASTING

There are basically two methods of determining whether a station from a particular area can be received at a given time via ionospheric reflection (sky wave). The more complicated method involves a mathematical or graphical calculation of the propagation path between two points, the simpler one merely involves learning to judge propagation conditions by the reception of various stations. The latter method is certainly the most useful for the hobbyist and is covered in detail. However, even the hobbyist should appreciate the more exact propagation forecasting methods possible.

Forecasting By Mathematical-Graphical Means

Although covered only in general terms here, the reader who wishes to learn more about this method should send for a listing of the publications on the subject from the Environmental Sciences Services Administration, U.S. Department of Commerce. They publish a number of periodic reports on calculated and observed ionospheric conditions, probably the best known being the monthly CRPI Ionospheric Predictions (15 cents a copy or $1.50 a year) and the Handbook for Use of CRPI Predictions (40 cents), both available from the U.S. Government Printing Office, Washington D.C.

Briefly, if you want to know whether a certain station can be heard at different times and on different frequencies, the following procedure would be used:

a. Find the great circle distance between transmitting and receiving points.

b. Depending upon the distance involved and the angle of antenna radiation, determine the refraction points or "control points" in the ionosphere.

c. From CRPI charts for the month, the maximum usable frequency at these control points is determined. The lowest maximum usable frequency, of course, determines the maximum usable frequency of the entire path.

d. From absorption index charts, which take into account seasonal and solar activity factors, the path attenuation is determined.

e. The transmitter power multiplied by its antenna gain is reduced by the path attenuation to give a graph of received field strength.

Whether the signal can actually be heard at the receiving point now depends upon the discrimination of the receiving

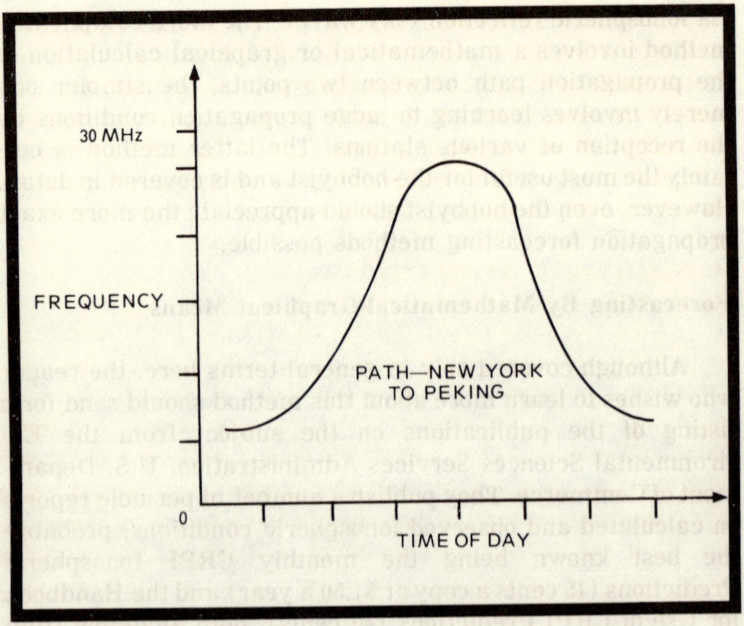

Fig. 1-6. At any given time of the day, there is a short-wave frequency best suited to reach any two points on earth. Such data is available in various handbooks, periodicals, etc.

antenna against atmospheric noise, the internal noise of the receiver, the local noise level at the receiving site and a correction factor for the type of modulation used.

Another simpler method of determining whether a propagation path is possible between two points more than 2500 miles apart, without regard to the transmitting or receiving equipment involved, is as follows:

a. On a great circle chart locate the "control points" at 1250 miles from each end of the path.

b. By reference to CRPI publications, notice the maximum usable frequency for each control point. The lower maximum usable frequency is taken as the path.

Plotting these maximum usable frequencies for various times of the day would produce a graph as shown in Fig. 1-6. Although relatively simple, many commercial and military services use such graphs to set up transmitting and receiving frequencies, trying to choose a frequency about 20 percent less than the maximum usable frequency. The receiving operator can, of course, use them also to great advantage when a choice of frequencies is available for the reception of a station. Any number of periodicals have done the work just described for the hobbyist and provide simple tabular monthly listings of which short-wave frequencies may be used at any time of the day to reach any point in the world.

FORECASTING BY RECEPTION

Since international broadcasting stations are interested in delivering the maximum possible signal, they choose their frequencies carefully from among those allotted to them. By studying their schedules of broadcasting hours to various parts of the world, they can take advantage of their frequency engineering efforts. For instance, if you were interested in general reception from the Far East in North America, studying the current broadcasting schedule of some major broadcaster from the area, say Radio Peking, would be very useful. The schedules would indicate during which parts of the day they beam broadcasts to North America and on which frequencies. During these same times and in the same frequency range, you could anticipate reception from a wide area of the Far East. The same procedure can be used for any other part of the world, of course. When you gain some experience using this technique, it is quickly possible for the hobbyist to judge reception or transmission possibilities on several portions of the short-wave spectrum in a few minutes.

There are some limitations with this technique, however, which should be understood. All stations are restricted to certain frequency assignments. The maximum usable frequency may be higher, for a given path, than a station can take advantage of in its transmission schedule. Also, transmission schedules must be developed some time in advance and retain a consistency in frequency usage in order to make it easier for the non-technical listener to find the station. Lastly, broadcasts to a given area are usually timed for hours when the largest audience can be expected. This is particularly true for international broadcasting stations which may, for instance, broadcast only half an hour per day in English to North America.

Nevertheless, by learning to check various station schedules and by receiving the stations themselves, you can save an immense amount of time compared to just random listening to check reception or transmission possibilities. Operators who really become proficient can make a reasonable estimate of how strong their own signal will be in a given area by judging the relative strengths of signals of different power levels from that area.

The transmissions from time and frequency standard stations maintained by various governments can also be used to great advantage to check propagation conditions. Besides just providing a signal to check, some of them broadcast information about propagation conditions. The next chapter tells how and where you can obtain information about these stations.

Frequency Bands and Assignments

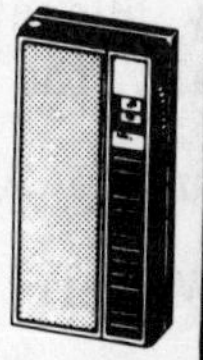

Chapter 2

Everyone is aware that certain government agencies license and control the use of various frequencies. But, it is often not appreciated that individual government licensing is part of a larger international effort to regulate the use of the radio spectrum. This may not be of too much concern on the VHF frequencies where only local communications are carried out. However, it is of particular importance on the short-wave bands even though it seems that far less order exists as you tune across them. Code stations are active in the middle of broadcasting bands, broadcasting stations in the middle of amateur bands, etc. Almost all these stations are operating legally—as far as licensing by their own government is concerned.

Technical as well as historical factors have determined the allocation of certain frequencies to the various users of short-wave spectrum space: broadcasters, maritime, land communications, amateurs, etc. And through international cooperation, there is a constant effort, utilizing all technical means, to accommodate an ever-expanding number of stations in the expandable portions of the frequency spectrum.

The international agency concerned with frequency allocation is the International Telecommunications Union (ITU), Geneva, Switzerland, which is a control agency of the U.N. By various treaties and conventions, the frequency spectrum has been divided into bands for various services. Individual governments determine the exact frequencies for their stations and impose technical controls. In the U.S., the agency responsible for non-government frequency control is the FCC, created by Congress with the Communications Act of 1934 (government frequency control is the responsibility of the Office of Emergency Planning).

The situation, however, is not so simple in practice. Due to technical considerations as well as pressures exerted by various countries at international conventions, the ITU has divided the world into three regions, each with slightly different frequency allocations in some cases for the same services. Generally, the regions are as follows:

Region I—Europe, Africa, USSR and Asia

Region II—Western Hemisphere

Region III—Oceania, Australia and part of Asia

It would be more confusing than useful to present the allocations for each region. Some of the differences in allocations for example, concern the amateur bands. In Region I, the 80 meter (3.5-4.0 MHz) band is shared by fixed and mobile commercial and government stations. Most countries allocate only 3.6 to 3.8 MHz for amateurs. Also in the Region I 40 meter (7.0-7.3 MHz) band, the frequencies from 7.0 to 7.1 MHz are allocated to amateurs. The rest of the band is a legal broadcasting band.

Besides the existence of different allocations in the various world regions, another problem is the registration of frequencies by broadcasters. Broadcasters (through the FCC for US stations) register their frequencies with the International Frequency Registration Board (part of the ITU). Disputes are generally settled in favor of the earliest registration. However, there is no real measure of whether or not a station or country really uses the frequencies they register and estimates are that 30 percent of the registered frequencies are really not fully utilized. Out of impatience with this situation, or for other reasons, some countries simply disregard completely the ITU frequency allocations. For instance, in the 7.0 to 7.1 MHz band, which is amateur in all regions, there are at least 10 major international broadcasters in operation, ranging from Radio Peking to Radio Iran.

In spite of all this, there is still a good deal of order on the HF bands and each service is allocated sufficient space throughout the HF spectrum so it has a choice of frequencies to compensate for changing propagation conditions.

FREQUENCY LISTINGS AND SCHEDULES FOR INTERNATIONAL SERVICES

When you consider that almost every country in the world licenses a multitude of transmitting stations for international and domestic broadcasting, as well as those for government and private uses, it becomes clear that to get any real pleasure out of using a receiving station, you should secure transmitting schedules for the types of stations you are interested in receiving. Also, by referring to program schedules and frequency listings, you can readily identify the stations heard, depending upon the language used. The latter is an extremely unreliable indication of station origin, though, since most international stations broadcast in a number of languages (Radio Moscow in at least 60, for instance).

There are a number of booklets and periodicals which give station and frequency listings for international broadcasting stations. Listed here are a number of publications and the areas they cover. Although some international stations have used the same frequencies for years (for example, West Germany on 5,960 kHz, England on 6,110 kHz, Moscow on 7,555 kHz and Peking on 9,480 kHz), specific frequencies are not listed here because most schedules change seasonally and any such listing would soon become out-dated.

Broadcast station listings are available from a great variety of sources, although most of them, except listeners' "heard" reports, are based on the publications of a few major organizations. Listings are published in a variety of time zone schedules, although GMT (Greenwich mean time) is probably the most common designation used. Clocks are available with indications for various world time zones, or perhaps a less expensive approach is the use of an extra clock set to indicate GMT since this is the only time zone, other than local time, of interest generally. Station listings also sometimes give the assigned call letters of the stations, although they are generally announced frequently only by domestic broadcasting stations and rarely, if ever, by international stations. Undoubtedly the largest listings of broadcast stations, and probably the most exhaustive, are those prepared by the International Telecommunications Union in Geneva, Switzerland. The major publications, which can be ordered directly from ITU, are listed below, with the cost shown in Swiss Francs. A free index of publications is also available.

Table of frequency allocations (map), 10 kHz to 4,000 kHz—14 Fr.

Map of coast stations—5 Fr.

List of broadcasting stations, 5,950 kHz to 26,100 kHz—31 Fr.

List of ship stations—7 Fr.

Annual HF broadcasting frequency list—62 Fr.

Quarterly HF broadcasting schedule—price on request.

List of fixed stations, international service—25 Fr.

International Frequency list, 9,995 - 28,000 kHz—80 Fr.

International Frequency list, 28 - 50 MHz—11 Fr.

Frequency plan for aeronautical mobile service—11 Fr.

List of low-power broadcasting stations, 41-230 MHz—7 Fr.

List of point-to-point radio telegraph channels—10 Fr.

Codes and abbreviations used in international telecommunication—7 Fr.

The English edition of the above lists should be requested. Publications can be obtained directly from the Secretary General, International Telecommunication Union, Geneva, Switzerland.

Although the ITU publications are quite detailed, to cover the world's stations requires assembling a set of books which together approaches the size of an unabridged dictionary. Therefore, some form of condensed listing is usually of more interest to most listeners.

For those interested in the domestic European broadcasting area, the European Broadcasting Union, Brussels, offers an interesting variety of publications. Their publications cover the European area including the Eastern European countries and European USSR. Some of their publications are:

a) Annual listing of long- and medium-frequency broadcasting stations by countries—50 Francs.

b) Annual broadcasting map of the European areas—25 Francs.

All publications can be ordered directly from the EBU, 32 Avenue Albert Lancaster, Brussels 18, Belgium. A free index of publications is also available.

The United States Government Printing Office, Washington, D.C., publishes a series of volumes which cover foreign HF broadcasting stations. Don't be misled by the title of "AM" stations. It doesn't refer to stations in the medium-wave AM broadcast band but to shortwave stations using amplitude modulation. Volume II is a particularly useful reference manual. Available at $2.50 per copy, the volumes are:

I—AM Stations by Country and City

II—AM Stations by Frequency

III—FM Stations

IV—TV Stations

There are a great number of magazines and booklets that list broadcast stations heard by readers or which give listings summarized from other publications. The value of some of these publications for the newcomer is debatable—for the price—as a source of detailed information. The only publication which is recommended is the World Radio-TV Handbook (World Radio-TV Handbook Ltd., Sundig 6, Hellerup, Denmark, and also distributed in the United States by Gilfer Associates, Box 239, Park Ridge, New Jersey 07028). Although written from a European viewpoint, its listing of program times, stations and frequencies of the major broadcasting organizations throughout the world is probably the best, up-to-date summary available. Also particularly valuable for QSL and correspondence purposes are the addresses listed for the broadcasting organizations of each country. Once you are more acquainted with broadcast reception, and if you want to pursue this aspect in greater detail, the bulletins of some of the DX clubs may be of interest. The completeness, accuracy and value of such publications vary greatly. Some of the clubs are:

a) American SWL Club, Watt Drive, Tallahassee, Florida

b) National Radio Club, Box 63, Kensington Station, Buffalo, N.Y.

c) Worldwide Monitors Radio Club, 782 S. James Rd., Columbus, Ohio

d) International Short-Wave Club, 100 Adams Garden, London S.E. 16, England

e) International Short-Wave League, 12 Gladwell Rd., London N. 8, England

f) Canadian International DX Club, 616 Stewart St., St. Charles, Manitoba, Canada

You can also request a copy of the current broadcasting schedule directly from an international broadcasting station. Most stations are very pleased to furnish such schedules and some are printed in quite attractive formats. Some stations also publish periodicals which are available for a nominal subscription fee. They give interesting information on program changes, station personalities, and technical changes. The only disadvantage in receiving program schedules from stations of strong propaganda agencies is that usually you will be sent only the schedules of broadcasts directed to your country. This can sometimes lead you to miss some very interesting listening. For instance, the difference in content and emphasis between the English language news and commentary programs broadcast by Radio Moscow to North America and to Africa is sometimes quite startling.

LISTINGS FOR COMMERCIAL, AMATEUR AND PUBLIC SAFETY STATIONS

Within the U.S., the FCC is the licensing authority for all non-commercial radio stations. What is generally not realized by the hobbyist is that the assignment of frequencies to FCC licensed, non-government stations, is a matter of public record. No public listing of government operated stations is available, nor would it always be practical, since block assignments may be made for a geographical area to a department of the government. However, the listing of non-government station assignments is available for public reference in the FCC Washington Office. The FCC itself does not reprint such listings for public sale but allows Cooper Trent Inc., 1130 19th St. N.W., Washington D.C., to reprint and sell such lists. The content of the lists is, of course, constantly changing, but currently the following lists of stations are available:

1. Broadcast Service List (AM, FM, TV, international)
2. Common Carrier List (International Fixed Public)
3. Experimental Service List
4. Industrial Service List (Special Services and Business Radio)
5. Land Transportation Service List
6. Marine Service List
7. Public Service List (Police)
8. Public Service List (Other Groups)
9. Class A Citizens

All lists contain information about the frequency, power, call sign assignments and geographical location of the stations in each class. Simply because some of the lists are hundreds of pages long, the cost becomes quite high, ranging from a few dollars to a hundred dollars. Inquiry should be made directly to Cooper-Trent for the current price of any listing.

A great number of smaller companies compile smaller selected lists of frequencies for specific types of stations in a certain geographical area such as the public safety stations around certain cities. These listings are usually smaller than those listed above and obviously are more modestly priced. You can locate these companies by scanning through the classified ad section in many of the monthly electronics magazines.

The best listing for U.S. amateur stations, as well as for most foreign amateurs, is that provided by the Radio Amateur Callbook (4844 W. Fullerton Ave., Chicago, Ill. 60639) which lists call letters and addresses of radio amateurs. Separate U.S. and foreign editions are available. The primary value of the book is as an address listing for QSL purposes.

For those interested in practicing morse code reception, either in preparation for an amateur license or simply to expand the ability to receive a greater variety of stations, many amateur stations send code practice transmissions. A listing of such stations can be obtained from the American Radio Relay League, Main St., Newington, Connecticut 06111.

LESSER KNOWN SERVICE LISTINGS

Another government publication that may be of interest is "Merchant Vessels of the United States," available from the Government Printing Office, Washington D.C. It gives the call letters assigned to United States merchant ships and would be useful for those who like to listen in on maritime CW activity. Specialized listings of weather, beacon and other public service government stations are available from agencies such as the Federal Aviation Agency, the Weather Bureau and Coast Guard. These agencies should be contacted directly for a listing of their current publications. Some other interesting listings are:

1. Newly Licensed Class D Citizens (Applications Review, 1217 Chillum Road, Hyattsville, Md. 20783). This is a weekly listing giving addresses and call signs of new Class D licenses.

2. Industrial Communications (Washington Radio Report, Ind., 1327 F. St., N.W., Washington D.C. 20004). This is a weekly listing of new stations in the industrial, land transportation and public safety radio services.

HOW TO READ, AND WHAT TO BELIEVE IN PUBLISHED SCHEDULES

Most schedules are quite straightforward in their presentation. The frequency, location, name or call sign of a station is given, possibly with its power (in kilowatts) and the area to which it directs its programming. The latter may be expressed directly (such as South America) or a zone number may be given. The zone number refers to the numbered zones of the world which the ITU has designated. This zone number designation will generally be found only in ITU publications and almost all of the publications include a small world map which shows the geographical coverage of the zones. Other listings may note only the bearing in degrees (referred to true north) to which their antenna is directed. Mixed schedules, which include all the stations in a particular frequency range regardless of whether they are voice broadcasting stations, radioteletype stations, etc., invariably designate the modulation type used by each station as well. This designation is a combination letter-number group such as A1 or A3. The meaning of these designations is shown in Fig. 2-1.

A0 - UNMODULATED SIGNAL

A1 - KEYED CODE SIGNAL (CW)

A2 - TONE MODULATED AND KEYED SIGNAL

A3 - AMPLITUDE MODULATED (AM), POSSIBLY SINGLE SIDE-BAND (SSB)

A4 - FACSIMILE

A5 - TELEVISION

F1 - FREQUENCY SHIFT KEYING (USUALLY RADIO TELETYPE)

F2 - FREQUENCY MODULATED TONE

F3 - FM PHONE

GMT - GREENWICH MEAN TIME (11 PM or 2300 GMT IS 6 PM or 1800 EST, ETC.)

UT - UNIVERSAL TIME (SAME AS GMT)

Fig. 2-1. Common abbreviations used in international station schedules.

The main difficulty with most schedules, particularly those for international services on the short-wave bands, is that they are out dated soon after publication. International broadcast stations usually divide the year into four broadcast "seasons" and change some of their operating frequencies each season. You can try to obtain schedules directly from the stations of interest, which is probably the best way to keep current on a station's frequency changes. Any number of periodicals also have large short-wave listeners' sections that list frequency changes and reader reports of unusual stations they have heard. Unfortunately, in some publications the latter tend to be more on the dramatic than on the useful side. It seems that many are trying to discover a spy station or report a station on a completely wrong frequency (which may be due to poor receiving equipment or not knowing how to use such equipment properly). But, there are many fine, serious SWL publications; you simply have to search them out.

A distinction also should be made between "registered" frequency listings and actual "monitored" frequency listings. Most of the frequency listings provided by the ITU, for instance, are registered listings. That is, a country submits to them a registration for usage of a certain frequency at a certain time from a given location within the registering country for service to a given area. Registration does not always mean the frequency will be used in the manner prescribed, if at all; however, it should mean this. Also, the locations shown within the country for a station are often incorrect for Communist countries. Monitored frequency listings, on the other hand, indicate that the station has actually been heard in the manner in which it is listed.

Receivers

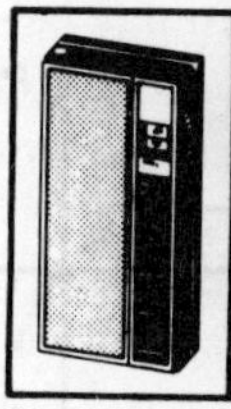

Chapter 3

Receiving equipment has undergone many amazing changes in recent years. The accelerated pace in technical development has been necessary due to the expanding demand for such equipment and because of the extremely crowded conditions on the short-wave and VHF bands.

However, receivers have also become more specialized and the qualities which modern receivers possess may not be suitable to all applications. For instance, stability requirements are very different for a receiver which is constantly tuned across a band in comparison to one which is used to continuously monitor one frequency. Chapter 4 presents a sampling of typical production receivers of various types and date of manufacture. Some hints on buying receivers are contained in Chapter 5 for the newcomer. However, this book is not intended as a buying guide. Space is not available to review all the receivers on the market nor would such a presentation be valuable for any length of time, since models change so rapidly. The reader can best concentrate on developing an understanding of the features which indicate the quality of receiver performance. With this knowledge, the evaluation of any current production receiver should become a much easier task.

HISTORICAL CIRCUITS

The basic purpose of any receiver is to select a desired signal and then to process that signal to extract the information it conveys. The simple diode receiver shown in Fig. 3-1 can remove the intelligence carried by an AM signal. The tuned circuit selects the station and the diode "extracts" the audio information from the carrier. The audio signal is then usually used to operate some electro-mechanical device such as headphones so the audio signal can be heard. The circuit

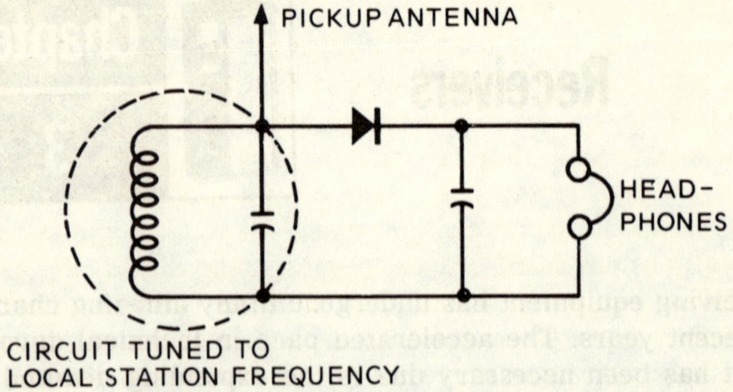

Fig. 3-1. Schematic of a simple diode receiver.

has a multitude of disadvantages, however, and is used only as a teaching aid today.

The regenerative type circuit and its variations, using either tubes or transistors, can be used to receive either AM or CW signals. A typical circuit is shown in Fig. 3-2. If the coil is adjusted so that the circuit does not oscillate, the AM signal is rectified by the diode action in the tube between grid and cathode and amplified by the normal triode action of the tube. For CW reception, the coil is adjusted so that the circuit oscillates at a frequency slightly removed from the incoming signal frequency so that the difference between the two frequencies is within the audible range and can directly operate headphones or a loudspeaker. The regenerative circuit, and particularly the super-regenerative variation, can be very sensitive receiving devices. However, the inherent disadvantages, such as poor stability, high noise and sensitivity of adjustment for optimum performance, have made them nearly obsolete today as practical receivers, except in all but the lowest priced applications.

Attempts have been made to improve the performance of such receivers by putting one or more stages of amplification before the regenerative stage. Isolating the regenerative stage in this manner and building up the signal before it reached the regenerative stage did improve performance. Today, however, this tuned radio frequency type of receiver, as it is called, has practically no application.

THE SUPERHET

The superheterodyne circuit, developed in 1916 by Edwin H. Armstrong, is the basis of almost all receivers today. The basic superheterodyne is shown in Fig. 3-3. The heart of the circuit is the mixer-oscillator combination which performs a frequency translating or "heterodyne" function. The incoming signal enters the mixer circuit along with a signal from the oscillator. The output signal taken from the mixer is basically a sum or difference frequency signal. As long as the oscillator frequency is varied so that the sum or difference frequency remains constant for each desired incoming signal frequency, the stages following the mixer need operate at only one intermediate frequency (IF). This greatly simplifies operation of the receiver and allows the IF stages to be relatively simple in design for maximum performance at one frequency. Whatever characteristics the IF stages have (such as restricting the bandwidth of the signals that pass through) will, therefore, be effective throughout the receiver's range.

An RF amplifier may be employed ahead of the mixer stage, since most simple mixer stages tend to be so noisy as to obscure very weak signals. The RF stage ahead of the mixer

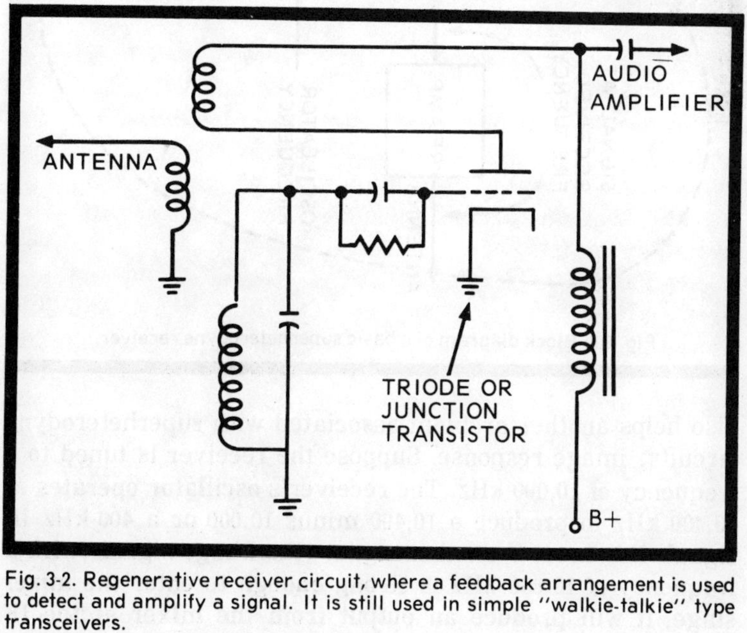

Fig. 3-2. Regenerative receiver circuit, where a feedback arrangement is used to detect and amplify a signal. It is still used in simple "walkie-talkie" type transceivers.

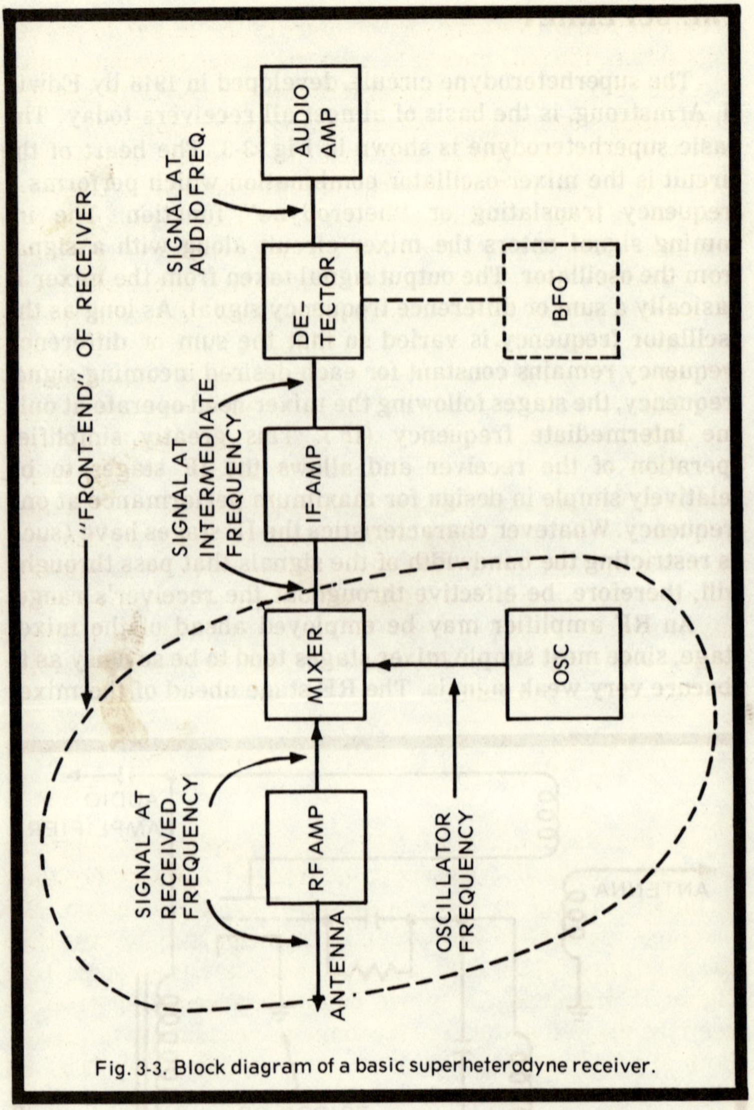

Fig. 3-3. Block diagram of a basic superheterodyne receiver.

also helps another problem associated with superheterodyne circuits: image response. Suppose the receiver is tuned to a frequency of 10,000 kHz. The receiver's oscillator operates at 10,400 kHz to produce a 10,400 minus 10,000 or a 400-kHz IF signal. However, if another signal (the image signal) with a frequency of 10,800 kHz is strong enough to enter the mixer stage, it will produce an output from the mixer at the IF frequency also. Nothing can be done to the mixer to prevent this action; the only cure is to keep the image frequency out of the mixer by having sufficient frequency selectivity ahead of that stage, such as the tuned circuits of an RF stage.

Although the image response tends to be the main interfering signal in most superheterodyne receivers, there is a number of other potential offenders. A strong signal at the IF frequency might enter the receiver and reach the IF stages, for example. Also, since the oscillator which supplies a signal to the mixer usually has harmonic content, these harmonics can mix with other signals arriving at the receiver and produce additional interfering signals at the IF. This situation can become complicated, with several such spurious signals occurring simultaneously. If the interfering signals are such that they produce two signals in the IF amplifier, differing in frequency by an audible rate, so-called "tweets" or "birdies" will be heard. Therefore, most receiver designs also concentrate on keeping the oscillator signal as pure as possible. When an RF amplifier stage is used ahead of the mixer, it can also generate harmonics of the input signal if the latter is strong enough to exceed the linear signal handling capability of that stage (overload).

MULTIPLE CONVERSION TECHNIQUES

If the desired signal frequency is relatively high as compared to the IF frequency, the image frequency becomes difficult to reject by circuits ahead of the mixer stage. For instance, the desired and image frequencies may be less than 100 kHz apart. The solution, of course, is either to go to a higher IF frequency or to use double conversion in order to keep the desired and image frequencies reasonably far apart. Both approaches are used in receiver design today. Fig. 3-4 shows how an additional mixer stage can be used. Some receiver designs use up to three stages of frequency translation. The disadvantage of multiple frequency translation is that one or more additional oscillator stages are required, and the oscillators themselves can generate spurious signals. The disadvantage of adopting a higher IF frequency is that the frequency selective devices, which the IF stages need to restrict bandwidth, are generally more costly and complicated.

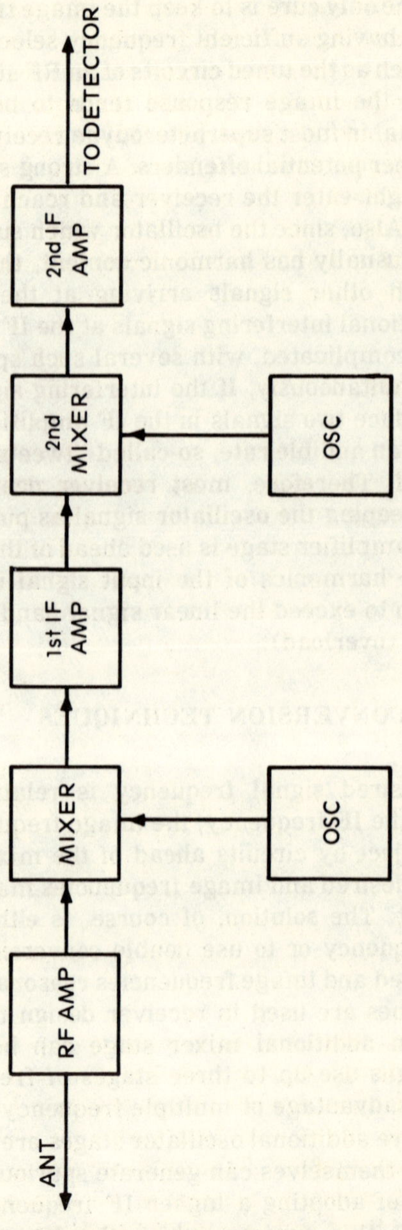

Fig. 3-4. Multiple conversion receivers have two or more mixer stages.

Although the bandwidth preceding the mixer stage(s) must be narrow enough to prevent image response, the major bandwidth restriction in most receivers takes place in the IF stages directly after the mixer stage. The mixer actually translates in frequency the band of frequencies contained in its input. The actual signal selection is done by tuned devices in the IF stages. This process need not be done in the IF stages. It can be done at the receiver input. But, it is generally more convenient to do it in the IF stages because they operate at one frequency and the selectivity device need not be retuned each time the input signal is changed.

FRONT-ENDS

The first three stages in a typical superhet receiver (Fig. 3-3), consisting of the RF amplifier stage, the mixer stage and the oscillator stage, are frequently referred to as the "front-end" of the receiver. This designation is used because these stages are usually the heart of the receiver. They process the weak incoming signal and initially amplify it. No matter how well the following stages perform, if the signal is lost at this point it cannot be recovered. Also, the oscillator stage, be it either tunable or of the fixed frequency type (crystal-controlled) determines the frequency which the receiver will accept and how constantly it will remain on that frequency. Again, no matter how good the subsequent receiver stages may be, including additional oscillator stages, the initial processing stages or "front-end" set the standard of performance for the entire receiver. Because of their importance, each of these stages is discussed in some detail in the following paragraph.

Naturally, when buying a commercial receiver you can't change the design of these stages. But understanding their function will allow you to better evaluate the "specs" of a receiver. Also, if you want to improve the performance of an existing receiver, one of the primary areas to look at—the area, in fact, where 90 percent of all improvements can be made—is the front end. So, don't become too discouraged by the detail included in the next few sections; rather, refer back to it as needed.

RF AMPLIFIER STAGE

The RF amplifier is probably the most important stage in a high performance receiver. It must have at least four important characteristics: enough gain so the signal level is raised above the noise level of the mixer stage; low inherent noise so incoming signals are not masked; enough selectivity so that at least image frequencies are rejected, and a sufficient dynamic range so that strong input signals will not cause the stage to produce spurious outputs.

Sensitivity is a term that is often loosely employed in reference to RF stages. Sensitivity is the input signal level required to produce some given output. For instance, for intelligible AM voice reception the output from a receiver must be, say, five times above the noise level at the output. The input signal level required to produce this output is the sensitivity. For an inexpensive broadcast band receiver, it might be 10 millivolts, while for a good short-wave receiver it may be 1 microvolt. The noise produced by the RF stage of the broadcast receiver is probably not very important since it would be low anyway compared to the relatively strong input signal needed to produce a usable output. However, in the case of the sensitive short-wave or VHF receiver, the noise produced by the RF stage is important since this noise level may approach the strength of the received signal itself.

The noise which an RF stage produces is caused by several factors all having to do with the electron flow in the passive elements (coils, resistors, etc.) as well as active devices (vacuum tube or solid-state device) which constitute the RF stage. Temperature, material irregularities and nonuniform electron emission are some factors which produce noise. In most passive and vacuum tube devices, the noise produced is distributed fairly uniformly over a wide frequency range. However, the noise voltage produced by vacuum tubes and solid-state devices is usually not so insignificant even over one band when very weak signals are considered.

A term used to measure the noise qualities of an RF amplifier is noise figure. This term compares the noise output of an amplifier to that of a theoretically noiseless amplifier which is said to have a 0 db noise figure. The term is independent of receiver bandwidth and is, therefore, a very useful one for general receiver performance comparison. The RF stage in a broadcast receiver may have a noise figure of 30 db or more, while in a good short-wave receiver it may be 3-6 db, and only 1-2 db for excellent UHF receivers. The noise figure becomes increasingly important as the reception frequency rises, since atmospheric noise decreases, as shown in Fig. 3-5, and RF stage noise is the cause of input signal masking.

Therefore, for general HF reception, you need a receiver with good sensitivity (a maximum of several microvolts or 10 to 20 db change in output level above noise) with as low a noise figure as possible so weak signals are not masked. Sensitivity and noise figures are mathematically related to each other and other RF stage characteristics, such as bandwidth, but in a complex fashion. The exact relationship is, however, not so important as the basic concept of having good sensitivity combined with a low noise figure.

The noise produced in the RF stage is extremely important because it determines the noise figure, basically, of the entire receiver. This can be proved mathematically. Since good orders of gain (from 20 to 100 times) are easily achieved in even simple RF stages, the noise figure of the RF stage overshadows that of any subsequent stages. If you

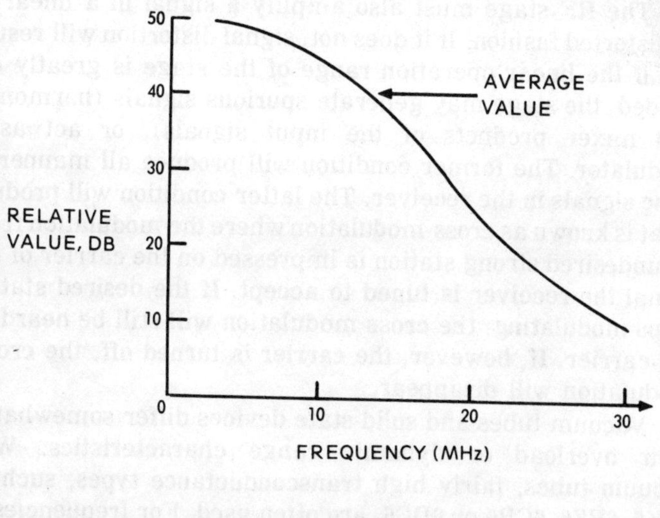

Fig. 3-5. Curve showing how atmospheric noise diminishes with frequency.

mathematically examine the simple input circuit of an RF stage, it can be shown that maximum signal transfer from the transmission line to the vacuum tube grid or transistor element does not occur under the same input circuit tuning conditions as those which produce a minimum noise figure. This is why the receiver input impedance may not match that of the transmission line. The input conditions for best noise figure lie between an open circuit and that which provides maximum signal transfer. Furthermore, the input circuit must provide some selectivity, at least enough to avoid having extraneous signals overload the RF stage and the possibility of image problems.

Generally, the tuning conditions for best performance of the input stage corresponds to that required for maximum signal transfer and not for best noise figure. Therefore, another conflict arises. To resolve these conflicts, many higher priced communications receivers resort to the use of two, three or even four RF amplifier stages in succession before the mixer stage. Others try to resolve the conflict by making the RF stage requirements less stringent by concentrating on low-noise mixer stages. Another approach is the use of multiple tuned circuits instead of a single-tuned circuit. Since you will encounter all sorts of variations in commercial equipment, obviously no one has found the ideal solution yet.

The RF stage must also amplify a signal in a linear or undistorted fashion. If it does not, signal distortion will result. Or, if the linear operation range of the stage is greatly exceeded, the stage may generate spurious signals (harmonics and mixer products of the input signals), or act as a modulator. The former condition will produce all manner of false signals in the receiver. The latter condition will produce what is known as cross-modulation where the modulation from an undesired strong station is impressed on the carrier of the signal the receiver is tuned to accept. If the desired station stops modulating, the cross modulation will still be heard on the carrier. If, however, the carrier is turned off, the cross-modulation will disappear.

Vacuum tubes and solid-state devices differ somewhat in their overload or dynamic range characteristics. With vacuum tubes, fairly high transconductance types, such as 6AK5, 6BZ6, 6CB6 or 6DC6, are often used. For frequencies in the upper HF and especially the VHF range, triodes such as 6BZ7, 6CW4 and 6AJ4 are quite popular. Mesa, silicon and field-effect transistors are most frequently used in transistorized receivers. Their dynamic range is not quite as good as vacuum tubes, however, although the newer MOSFETs approach them closely. Some designs, realizing this, use a combination of tubes for the RF stages and transistors for the following stages. Other designs accept the somewhat poorer dynamic range of the transistor and build in an attenuator directly before the RF stage so the operator can manually reduce the overall signal input level if the RF stage appears to be overloading.

The latter solution is by no means ideal since under very crowded and noisy HF reception conditions, cross-modulation effects are not always obvious even to an experienced operator. Such attenuators naturally reduce receiver sensitivity also—a noted disadvantage if set to accommodate the strongest station heard when tuning across a group of weak signals. There are some circuits that have exceptionally good overload characteristics, using both vacuum tubes and solid-state devices, that can perform mixing as well as amplification functions. Two typical RF amplifiers using tubes and transistors are shown in Fig. 3-6.

MIXER AND CONVERTER STAGES

A mixer can be any nonlinear device, even a diode. If the input and oscillator signals are applied to a diode, the resultant output is a complicated combination of the basic input and oscillator signals, their harmonics and combinations thereof. In normal HF receiver practice, only the sum and difference frequencies are of interest. In fact, usually only one of these frequencies is used, the other being rejected by a filter in the output of the mixer stage.

The diode mixer is not generally used in HF receivers because it provides no conversion gain (amplification of the input signal level) nor isolation to prevent oscillator signal feedback into the output circuit of the RF amplifier (and thence to the antenna to possibly radiate and cause interference to other receivers in the area).

Triode mixers are used frequently in HF receivers. The single-ended types are somewhat subject to a "pulling" effect; that is, a change in the input signal level will change the load on the oscillator and possibly its frequency, depending upon the oscillator design. The cathode coupled circuit ef-

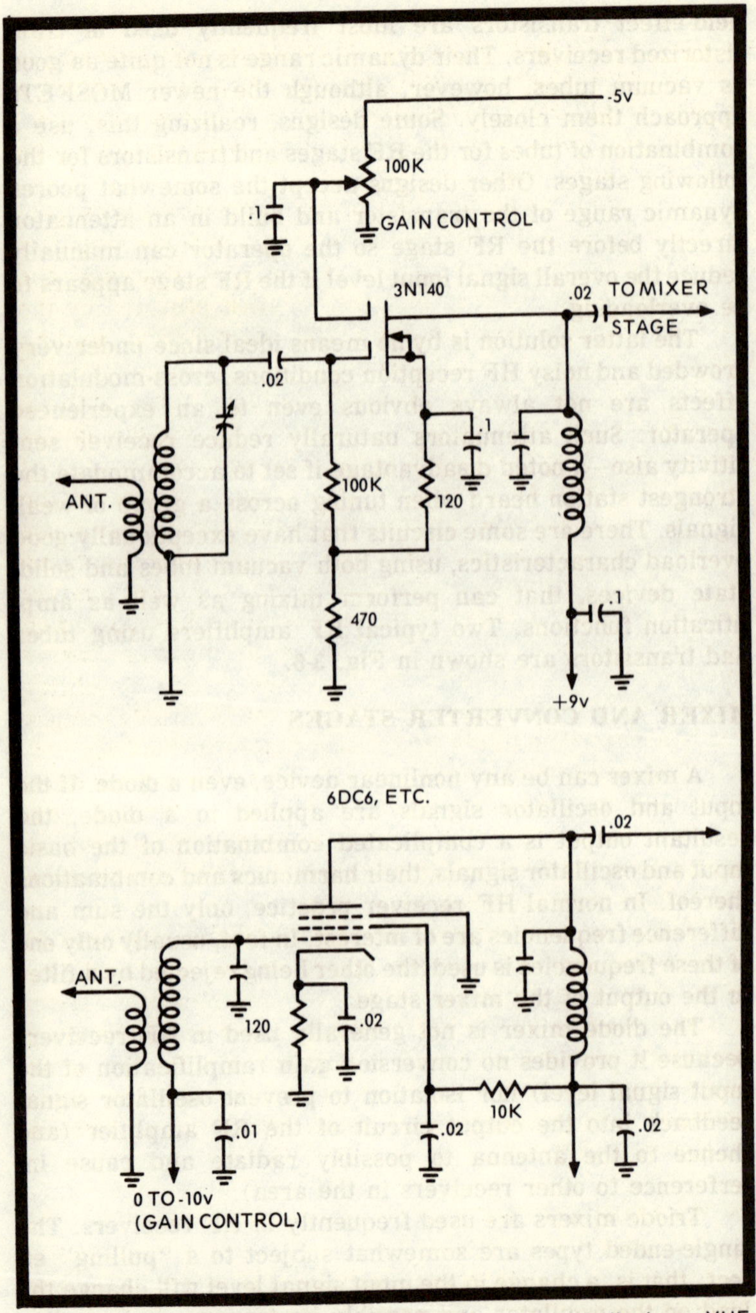

Fig. 3-6. Typical RF amplifier circuits designed for use up to the lower VHF range.

Fig. 3-7. Cont'd.

development of multi-channel transmission systems have demanded that receivers possess high selectivity.

The frequency of an oscillator is affected by a change in the voltages applied, temperature, humidity, mechanical stability, etc. In practice, temperature and voltage variations most affect frequency stability. Various means are employed to produce a constant environment for an oscillator: sturdy, "battleship" type mechanical construction, regulation of applied voltages, capacitors immune to temperature changes or which change with the temperature in an inverse manner to cancel the frequency change caused by temperature effects on other components which carry the frequency in an opposite direction, heat ovens to provide a constant temperature, etc.

The frequency stability of a receiver is often stated in terms of the frequency variation that occurs over a given period of time (assuming a constant external environment) or over a given range of temperature (usually freezing to about 150 degrees F) or power line voltage variation (usually plus or minus 10 percent). The frequency stability (for time, temperature or voltage change) may be stated as being plus or minus a maximum of so many Hertz or kiloHertz, percentage of frequency or part of frequency. However, a statement that a receiver has a drift of 100 or 500 Hz per hour at 30 MHz sounds good but may not mean too much. If the receiver is used for continuous tuning of stations, where it will rest on each frequency for relatively short periods, 500 Hz per hour-stability may be quite satisfactory for normal CW, AM and SSB reception. If, however, the receiver is used to listen to one frequency only, the tuning may change so much in the course of a day that the station is lost. (Remember no **maximum** absolute drift was stated; only 100 or 500 Hz per hour.)

Tunable HF oscillators of reasonably good construction may have temperature stabilities at 30 MHz of from plus or minus 3,000 Hz to a few hundred Hz. Simple crystal-controlled oscillators or tunable oscillators referenced to a crystal-oscillator may have stabilities of from about plus or minus 100 Hz to a few Hz. Crystal oscillators in temperature ovens may have stabilities of a small fraction of a Hz.

Stability should not be confused with frequency calibration, readout resolution or tolerance. For instance, a 10-MHz crystal may have a tolerance of plus or minus .01 percent (1 kHz). It indicates that the actual crystal frequency may be any value within 1 kHz of the stated frequency; the stability will depend on the conditions under which the crystal is used.

Typical continuously tunable or crystal-controlled oscillator circuits are shown in Fig. 3-8. In a simple superheterodyne covering 500 kHz to 30 MHz, the tuning range is usually divided into several bands. For each band there are separate frequency-determining portions of the oscillator circuit and a bandswitch introduces the appropriate portion for each band.

In a double superheterodyne circuit designed to continuously tune a band of frequencies, one of the oscillators is usually crystal-controlled and the other variable, although there is one interesting scheme used on VHF sometimes to obtain double superheterodyne action with only one variable oscillator. When one crystal-controlled and one variable oscillator are used, the crystal-controlled oscillator design depends on several considerations. If the first oscillator is made variable, the output of the first mixer will always be the same (as with a single superheterodyne); therefore, the second oscillator needs only one crystal to translate the signal to the IF frequency in the second mixer.

If the receiver covers only small portions of the HF range, the variable oscillator can be made reasonably stable and accurate. However, coil switching is still necessary. However, if the receiver covers the entire HF range up to 30 MHz, the variable oscillator must operate over several megaHertz at least, depending upon how many bandswitch positions are used to cover the entire range. Making a variable oscillator operate over such a range entails some compromise in stability. Many receivers still use this scheme and, if well constructed, exhibit stability characteristics quite adequate for general listing purposes, particularly for AM reception.

However, for reception purposes where greater stability is desired, the usual scheme is to make the first oscillator crystal-controlled and the second oscillator variable. If the crystals used in the first oscillator are properly chosen, the output of the first mixer will always cover the same range for any input frequency range. The variable oscillator then has to cover only one frequency range for all the receiver input ranges, thereby eliminating the need for oscillator coil switching. Under these circumstances, special care can be given to the construction of such an oscillator for maximum stability and linearity.

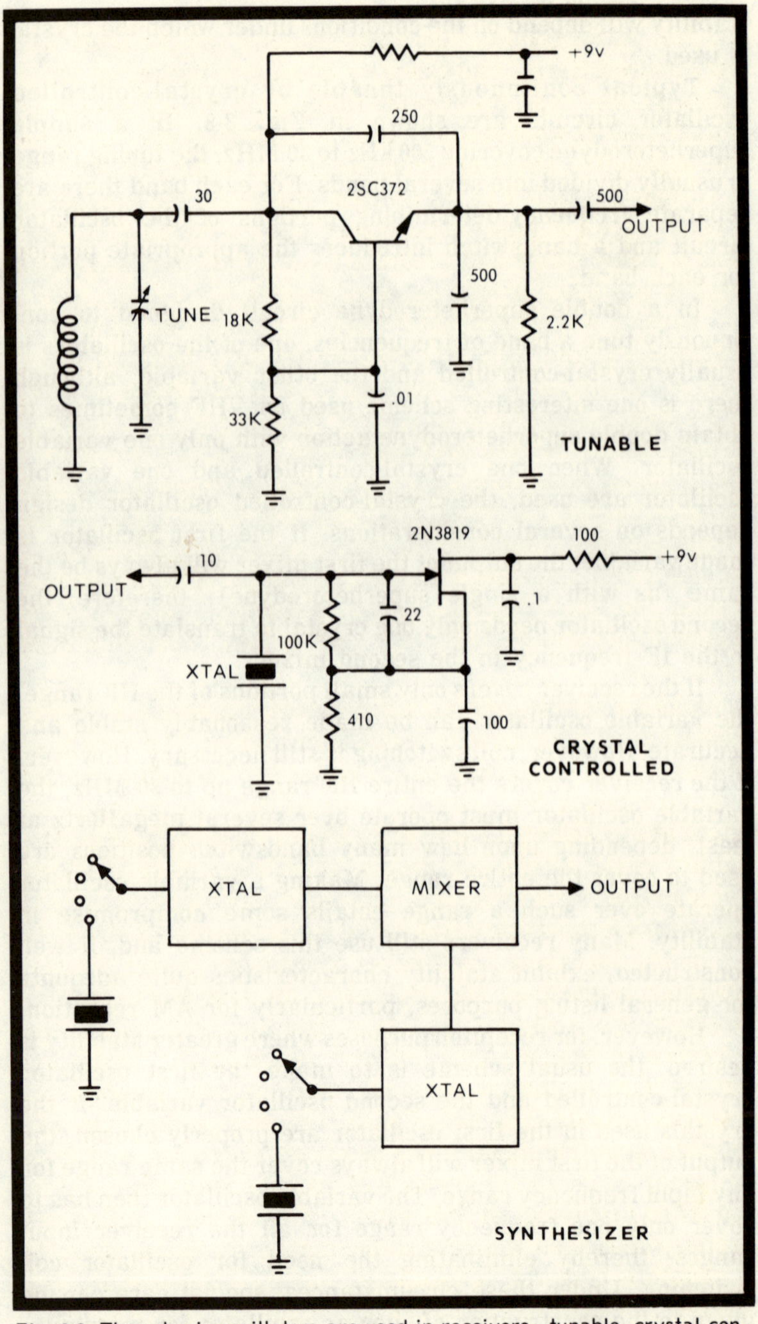

Fig. 3-8. Three basic oscillators are used in receivers—tunable, crystal controlled and synthesizer.

The maximum range which such an oscillator can cover and still retain good temperature stability (on the order of a few hundred Hertz) and linearity is usually limited to 500 or 1,000 kHz. However, to cover the range from the AM broadcast band to 30 MHz would require from 30 to 60 crystals. This is not true in practice, however, since harmonics of the fundamental of various crystals can be utilized. However, the number of crystals required plus the necessary quality (since each must be of a highly stable and accurate type) limits the use of such circuits to higher priced receiver models.

Various ways have been tried to reduce the number of crystals required. One involves the use of a frequency synthesizer. Essentially, it is just two crystal oscillators working into a mixer stage. If it is desired to cover a small band of frequencies in discrete frequency steps, such a circuit is very useful for crystal control of receivers or transmitters, for instance, on the CB band. By using the combination of the frequencies resulting from the mixer action, only a fraction of the number of crystals is needed as compared to a separate crystal for each desired output frequency. However, the disadvantage still remains that each crystal must be a very stable type.

Another synthesizer variation employs a group of crystal oscillators all referenced to one master crystal. The outputs of all the crystal oscillators are combined to form the desired output frequency. Such circuitry can produce, for instance, an output at any selected 1-kHz point from 1 MHz to 75 MHz. The stability of the output is as good as 75 MHz at 1 MHz. If a very stable master crystal is used, this could mean an output at 75 MHz with a temperature stability of a fraction of a Hertz.

In the foregoing synthesizer circuit, a master crystal oscillator is used as the reference for frequency control. However, it is also possible to use the received signal as the reference to produce another form of automatic frequency control. This method is normally used only for commercial overseas circuits. The use of such frequency control circuits depends upon the signal being received, of course. A shift in frequency of 100 to 150 cycles will produce noticeable distortion when receiving an SSB signal. If frequency synthesizer circuits referenced to a very stable master oscillator are used to generate both the transmitted signal and to control the receiver, the drift will be so negligible that no other frequency control is necessary. In SSB or multiple-channel service this means that no carrier for frequency control purposes is needed in the receiver. However, when the transmitted signal is not so stable, frequency control is still necessary even in a receiver where the tuning is referenced to a stable master oscillator. Another instance where frequency control may be necessary, even though both transmitter and receiver oscillators are extremely stable, is a case where the doppler shift must be taken into account, as with signals from man-made satellites.

TUNING AND FREQUENCY CONTROL SYSTEMS

In a crystal-controlled receiver the switch which selects different oscillator crystals is usually accompanied by a dial with markings for each frequency (or just channel numbers might be indicated as for a CB or marine receiver). Tunable receivers usually have a dial of some sort with an indicator linked to indicate the setting of the variable oscillator in the receiver, since the oscillator must always have a definite relationship to the incoming signal. A variable oscillator can be tuned by means of a variable capacitor or inductor or electrically by a varactor (a special diode which replaces a variable capacitor). In any of these cases, an indicating scale can be mechanically linked to the device that is moved to change the oscillator frequency. If the oscillator output is linear, the scale divisions on the frequency indicating dial will be uniformly spaced. If the output is not linear, the scale divisions for some given frequency coverage (10 kHz, for instance) might be spaced one-sixteenth inch apart on one part of the scale and ¼ inch on another part.

If a variable capacitor is used to tune an oscillator, one rotation of the capacitor might vary the frequency 1 MHz. This would, of course, be much too unhandy to use since the tuning would sweep through a high-frequency band with almost half a turn of a knob. So, some means of mechanical linkage or a separate "bandspread" capacitor with a fraction of the capacity of the main capacitor is used to reduce the tuning rate. Tuning rates as low as 10 kHz per knob revolution have been found to be very desirable for tuning in SSB and teleprinter signals. Larger tuning rates, up to 100 kHz per knob revolution, can be used for AM reception. But rates greater than this generally are undesirable under crowded band conditions, particularly below 30 MHz.

Frequency readout depends upon the spacing between frequency markings. For general HF reception, frequency markings at least every 10 kHz are desirable. If these markings are spaced ⅛ inch to ¼ inch, visual interpolation to a few kHz is possible. If the receiver has an oscillator that produces marker frequencies every 500 Hz or 1 kHz, the electrical interpolation might be 1 kHz. Many more expensive and stable receivers have frequency markings every 1 kHz and visual interpolation is usually possible to about 200 Hz.

Backlash and resetability are two problems that arise because of the "play" in the mechanical linkages in a tuning system. Backlash simply refers to how much the tuning knob can be rotated, back and forth, without actually changing the receiver frequency. Resetability refers to the capability of tuning to the same frequency at the same dial marking after turning the dial through the entire range. Poor backlash and resetability usually occur in receivers after they have been used for some time. Normally, backlash and resetability limits should not be greater than the visual frequency interpolation possible.

Rather than have a pointer move over a scale, rotation of the tuning knob in some receivers activates a turns or Veeder-Root counter. If the oscillator output is linear, the turns counter can be calibrated directly in frequency. Some receivers use such counters with up to six counter wheels so that the frequency can be read on the counter display directly to 100 Hz. Other designs use only a digital wheel display to indicate the MHz range to which the receiver is set and a conventional dial and pointer for frequency readout within the MHz range. However, being a mechanical device, the Veeder-Root counter also has the problems of backlash and resetability.

One frequency readout method which eliminates most of the problems of the above methods is an electronic frequency counter. With the development of micro-circuits, such counters can be made small enough to fit into most receiver packages. But if such a counter is connected to measure the frequency of the variable oscillator in a receiver, the count will differ from the actual receiver frequency by the IF frequency. Receivers using this method usually are designed with an IF frequency of some even number such as 1 MHz. Therefore, by presetting the MHz count to plus or minus the IF frequency, as appropriate, the counter will display the received frequency directly. The display rate is made high enough so that the display appears to be continuous. Such frequency display units can be made the size of a small text book, with the only parts normally needing replacement being the neon number display tubes. The accuracy available, with readings down to 10 Hz, if desired, far exceeds that available with any mechanical arrangement. Backlash may still exist in the gearing between the tuning knob and the oscillator variable element. However, resetability is almost perfect, since it is equivalent to the accuracy of the counter.

IF AMPLIFIERS

The IF amplifier stages in a receiver serve two basic functions: first, to provide enough gain for operation of the detector stage and second, signal selectivity. Even though the RF amplifier and mixer stage(s) amplify the signal, additional gain is still necessary because the signal at the mixer output is still only several hundred microvolts. Since the detector usually requires a volt or more of signal for proper operation, the IF amplifiers must provide the needed signal boost. It should be noted that no amount of IF amplification will help if a signal has already been masked by noise in the front-end (RF amplifier and mixer) stages. Additional RF amplification will help if the present IF amplification is insufficient to give a weak signal sufficient boost to operate a detector stage. This point is often neglected when older receivers are rejuvenated by adding IF stages, when really the problem is a poor front end.

The typical IF amplifier stage looks very much like the RF amplifier stage in Fig. 3-6. It has tuned input and output circuits, but these circuits are fix tuned to one frequency (455 kHz, for instance). A number of such stages may be used following each other, and, besides the tuned input and output circuits in each stage (so-called IF transformers), they may contain other devices which determine the "selectivity" or the "bandwidth" of the IF amplifier chain. These terms should be thoroughly understood since they are quoted almost anytime you read about a receiver. The selectivity is a measure of the frequency response of the IF amplifier as shown in Fig. 3-9. The bandwidth (in kHz) is usually referenced to some amplification point such as -3 db or -6 db.

The needed selectivity depends upon the nature of the signal being received (Chapter 7). No more bandwidth than necessary to accommodate the signal should be used to avoid interference. However, when tuning across a group of stations, it is frequently useful to use a wider bandwidth in order to more quickly survey the transmissions present. Therefore, most receivers contain some means of controlling the bandwdith from perhaps 12 kHz for AM broadcast or multiple-channel transmissions to 100 or 200 Hz for keyed CW transmissions. The amount of selectivity which a receiver can tolerate is also dependent on the oscillator stability in the receiver. For instance, if the IF amplifiers provide a 200-Hz bandwidth, but the receiver oscillator drifts several hundred Hertz in the course of an hour, the operator is going to be bothered by the need to frequently retune the receiver.

FILTERS

Although all the tuned circuits in an IF amplifier provide some selectivity, any special device used to provide increased selectivity is usually installed in the first IF amplifier stage.

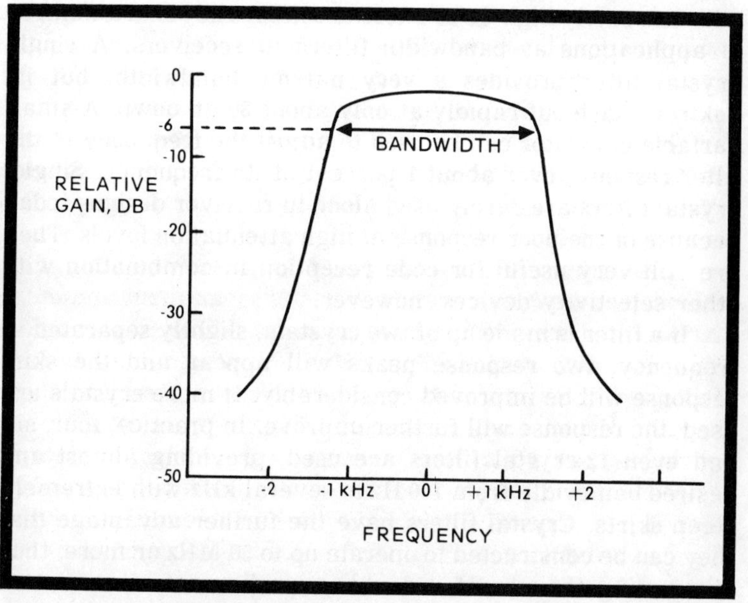

Fig. 3-9. Selectivity curve for an IF amplifier. The zero frequency shown represents the center frequency of the IF.

This is done simply to restrict the bandwidth as early as possible so that the succeeding IF stages are not overloaded by undesired signals. Some of the more common selectivity devices are described below.

a. Multiple IF transformers can be used; two transformers are connected back to back between each IF stage. Windings can be switched in or out to vary the bandwidth. The expense of special transformers and the need for switching the windings on several transformers simultaneously has all but outdated this method.

b. The Q multiplier endeavors to enhance the selectivity introduced by a single tuned circuit by using a feedback arrangement so the loop gain effectively multiples the Q of the coil (Q being a measure of the selectivity of a single tuned circuit). Thus, a coil with a Q of 100 and a circuit with a gain of 40 will simulate a coil with a Q of 4,000. If too much feedback is used, the stage will oscillate. Such circuits are generally found only in low-frequency IF systems, up to about 500 kHz, because of the difficulty of building simple but stable feedback circuits at higher IFs.

c. Quartz crystals can be analyzed as fixed-tuned circuits with extremely high Qs of 10,000 to 50,000. They find a number of applications as bandwidth filters in receivers. A single crystal filter provides a very narrow bandwidth, but its "skirts" flare out rapidly at only about 30 db down. A small variable capacitor can be used to adjust the frequency of the filter response over about 1 percent of its frequency. Single-crystal filters are rarely used alone in receiver designs today because of the poor response at high attenuation levels. They are still very useful for code reception in combination with other selectivity devices, however.

If a filter is made up of two crystals, slightly separated in frequency, two response peaks will appear and the skirt response will be improved considerably. If more crystals are used, the response will further improve. In practice, four, six and even 12-crystal filters are used, providing almost any desired bandwidth from 100 Hz to several kHz with extremely steep skirts. Crystal filters have the further advantage that they can be constructed to operate up to 50 MHz or more, thus eliminating the need for double superheterodyne operation where the purpose of the second frequency translation is to provide an IF low enough for an effective bandwidth filter.

d. Ceramic filters, or transfilters as they are also called, are also piezoelectric devices, like the quartz crystal, but they use a ceramic element. If enough elements are incorporated, the response is essentially that of a crystal filter with a similar number of elements. Ceramic filters find special applications in transistor circuits, since the low input and output impedances allow them to be used directly as the coupling element between stages. The main disadvantage is that they are limited in frequency to about 10 MHz.

e. Mechanical filters are electromechanical devices where electric energy is used to mechanically oscillate metallic elements (usually discs) and reconvert the mechanical oscillations into the electrical form. The metallic elements can be visualized as individual tuned circuits. Seven or more of these discs can be assembled in a small enclosure of a few cubic inches as compared to the large enclosure needed for the equivalent number of capacitor-inductor elements. The filters are commercially available in bandwidths of 500 Hz to 8 kHz. However, they are limited by material and constructional restraints to frequencies of 500 to 1000 kHz.

DETECTOR CIRCUITS

The detector circuit is the point at which the information contained in the radio signal is finally extracted or "detected." For an AM signal, the information is contained in amplitude variations of the signal and the detector is usually only a simple diode circuit that essentially rectifies the radio signal to obtain the audio frequency variations modulating the wave. It is sometimes also referred to as an envelope detector.

To detect a keyed signal which is not modulated (that is, the transmitter is just turned on and off), a so-called BFO or beat frequency oscillator is required. The BFO signal "beats" with the keyed signal in the detector. For instance, if a signal came through an IF at exactly 455 kHz, it would not be audible. If, however, it is beat or mixed with a BFO signal of 456 kHz, the result is an audible 1 kHz (1,000 cycle) signal.

SSB transmissions are essentially detected the same as keyed signals. Some form of BFO is needed to work along with the detector stage. The detector stage is usually a bit more elaborate than that used for CW reception and a type of circuit called a product detector is frequently used.

For FM signals, the intelligence is contained in the frequency variations of the signal, so the detector stage must be sensitive to frequency and not to amplitude changes in the IF signal. A number of circuits are available for this purpose, but because they do discriminate between frequency changes in the signal, they are referred to as "discriminators" in FM receivers rather than detectors.

The detector stage in almost all receivers also provides some measure of automatic gain control; that is, in detecting the incoming signal, its amplitude is also sensed and a feedback signal developed which regulates the gain of the RF amplifier stage and the IF amplifier stages. Thus, the output level of the receiver is held fairly constant for large variations in the strength of the input signal. Such systems are not completely successful and the operator usually still has to do some manual adjustment anyway. For some types of reception, it may be desired to disable this feature and shortwave receivers, especially, usually have a switch to disable the automatic gain control or automatic volume control (AVC) circuit when desired.

The usual "S meter" circuit is tied in with the automatic gain control circuit. It may indicate directly the level of the control signal, which is related to the strength of the input signal, or it may measure the current flow in some IF stage which is being gain regulated.

AUDIO SYSTEMS

Generally, the output of a detector stage is near a few volts at most. The audio amplifier stages provide enough power to drive some conversion device such as a loudspeaker or headphones. Power outputs range from a few hundred milliwatts to a few watts. Harmonic distortions of 5 to 10 percent are acceptable for general reception purposes. Such amplifiers are conventional in design. Sometimes the response is tailored for best response in the broad 300 to 3,000 Hz range, especially for speech reception, by using small coupling capacitors (.001 mfd) between stages and bypass capacitors to reduce the higher frequency response.

Providing any great degree of selectivity within the audio portion of a receiver is difficult because of the need for high-Q components at audio frequencies. It is also undesirable because interference signals can overload previous stages and control AGC action. Therefore, few receivers are built today with audio selectivity, although it still remains a simple way to possibly improve the performance of an existing receiver. Circuits are used to peak the audio response at a particular frequency. The sharpness of any of these filters depends upon the Q of the components used. "Range" filters, designed to sharply reject or pass 1020 Hz, are frequently sold on surplus markets and used as CW audio filters.

Sometimes, the audio section of the receiver part of a combined transmitter-receiver is "beefed-up" in power so it can also act as the modulator for an AM transmitter. It may also have provisions for use as a paging amplifier. In the latter case, the input to the audio amplifier is switched between the output of the detector stage and a microphone.

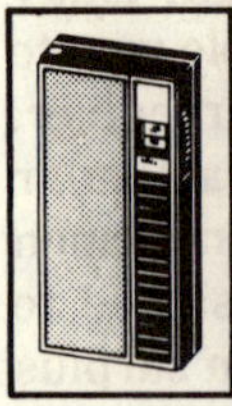

Chapter 4
Survey of Receivers

Receivers can be classified in a number of ways—by price, size, general application, etc. The classifications used in this chapter are arbitrary and intended only to point out some of the great variety of receiver designs in use. Although various manufactured receivers are shown for illustrative purposes, remember that they are only a small sampling of those currently available. Manufacturers are constantly bringing out new models with improved features.

The receiver classifications used here are "general coverage," single-band and fixed frequency, and receivers with extra features and those designed for specialized applications. Most of these terms should be clear in meaning. General coverage receivers refer to those designed to operate over a broad frequency range and are often referred to as multi-band receivers also. A typical receiver might cover from 550 kHz to 30 MHz in anywhere from 4 to 20 different segments. Less expensive receivers generally use fewer segments. For instance, the first segment or "band" position might cover just the AM broadcast band (550 kHz to 1600 kHz), with succeeding bands covering 1.6 to 6 MHz, 6 MHz to 15 MHz and 15 MHz to 30 MHz. Less expensive receivers of this type may also have no RF stage or BFO to receive CW signals. More expensive receivers incorporate these features as well as special bandwidth filters for AM, SSB or CW, noise eliminating devices, etc.

"Single-band and fixed-frequency receivers" refers, in the former case, to a receiver that is tunable over only one band and in the latter case to a receiver always locked on one channel by means of a crystal-controlled oscillator.

"Receivers with extra features" refers to those types which incorporate a mixture of features normally associated with, perhaps, a general coverage receiver or fixed-frequency receiver. Or, they may automate to some degree the tuning that is normally done manually.

"Specialized receivers" include mainly commercial receivers. Most are far too expensive for the hobbyist, but nonetheless they provide an interesting insight into how sophisticated the development of receiving equipment has become.

RECEIVER CONTROLS

The number and type of controls which a receiver has depends on its usage and the type of operator who will be using it. An FM receiver for public safety use may likely have only an on-off switch and a volume control since it is designed for unskilled operator use. A multiple receiver setup for long-distance HF multiple-channel reception will contain dozens of controls to permit adjustment of each major circuit function. Probably the radio hobbyist will derive the most useful information by a review of the control functions common to general coverage receivers:

Level controls. These usually consist of the RF level or gain control and another potentiometer in the input to the audio amplifier section called an audio gain control. Some designs have a variable attenuator in the antenna input circuit to the RF stage called the RF gain control. Although useful in eliminating overload in transistor RF stages, it may introduce noise, particularly as the potentiometer becomes worn. Other designs provide a switched attenuator, which may be cut out when overload is not a problem in conjunction with an RF gain control. This latter arrangement is probably preferable. An additional level control found in some receivers is an IF gain control, although it is common only in designs where excessive IF gain would affect the linear response characteristics of the circuit or a selectivity element in the IF.

Frequency selection controls usually provide a means of tuning the variable oscillator, RF input stages and possibly the tuned circuits in an intermediate amplifier between two frequency mixing stages. Often these controls are ganged mechanically or electrically so only one control is required. In other designs, they are not ganged. The tuning control for the RF amplifier stage, then usually called preselector tuning, has to be retuned for every frequency excursion of about 100 kHz or more. Other designs which use ganged controls may still have a control called antenna tuning. However, in this case, the control is a vernier across the ganged RF tuning, which allows a better match between the input circuit to the antenna. With an antenna system having a small impedance excursion with frequency change, this control will require almost no adjustment. If a receiver is to be tuned frequently over a wide range, a ganged arrangement is certainly preferable.

Bandspread usually involves a ganged tuning capacitor in parallel with the main tuning capacitor. A selected frequency range on the main tuning scale may then be effectively spread out over more turns of the bandspread tuning knob, thus allowing a separate bandspread scale to be used. In effect, however, it is just a means of requiring more turns of a knob to cover a frequency range. It cannot improve selectivity, although this effect is sometimes believed because stations are easier to tune.

A vernier or calibrate control is usually only a fine-tuning device across the oscillator tuning. It is not ganged to the main tuning control and it varies the oscillator frequency from 1 to 5 kHz to set the oscillator to correspond with the frequency scale markings. Another method used to accomplish the same purpose is a mechanically movable dial marker (pointer).

Bandswitch controls usually connect the appropriate RF, mixer and oscillator coils for the desired band. Certain switch positions also bypass various mixer stages in multiple-conversion receivers when they are not needed for a specific band.

Mode selection controls select the required detector for the type of modulation being received. If a crystal-controlled BFO is used for SSB, these crystals also must be switched. Other designs use a variable BFO which must be tuned for SSB or CW. The mode selector may also switch in different IF selectivities for different modes (½ or 1 kHz for CW, 3 kHz for SSB, etc.), but again separate controls for these functions are desirable.

Selectivity controls switch the filters used in the IF portion of the receiver. Generally, such controls are marked in bandwidth. When separate filters are not used, the control will usually continuously vary the setting of a single bandwidth restriction device. Passband tuning refers to a control which varies the center frequency of a selective device without changing its bandwidth. A phasing control, which has much

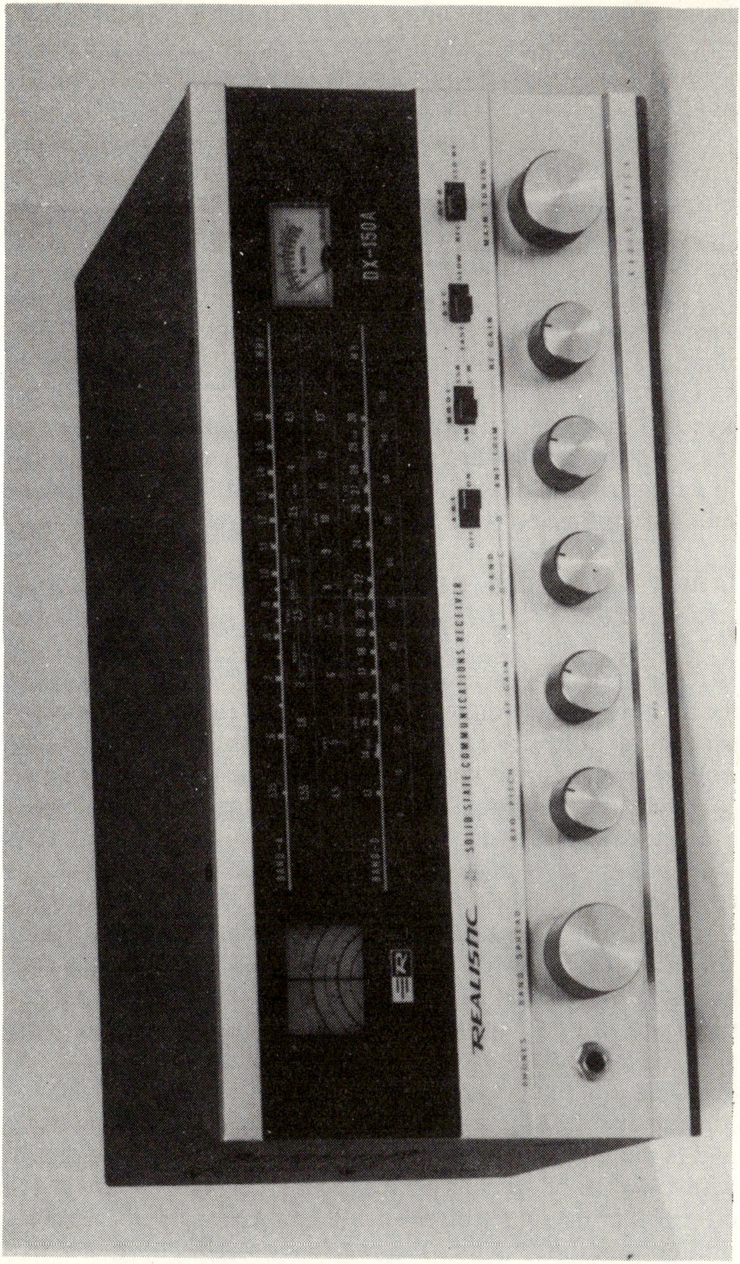

Fig. 4-1. The Realistic DX150A covers the entire 30-kHz to 30-MHz range in four bands.

the same action, is sometimes used with single-crystal filters. Notch frequency and notch depth controls tune a bridged-T network to the frequency to be rejected and determine the level of rejection.

Miscellaneous controls might include a noise suppression circuit on-off switch and threshold control, a calibrator switch to turn on a test oscillator for level or frequency scale alignment, a power switch and a standby switch to disable the receiver without turning the power off.

Various schemes have been tried for combining and placing controls. There is no fast rule for judging what combinations or placements are best. Overcombination certainly reduces the versatility of adjustments, but it simplifies panel layout. Placement of controls usually satisfies right-handed operators by having the frequency tuning knobs in the center or on the right side, while level and selectivity controls are situated on the left side.

LOW-TO-MEDIUM PRICED GENERAL COVERAGE RECEIVERS

The variety of receivers available for general coverage of the high-frequency range is almost unlimited. Those described here merely illustrate a few typical receivers, and it would be well to underline the word "typical." The average hobbyist, particularly when starting out, would do well to stay with conventionally designed receivers until he is familiar with the functions of a receiver and can better analyze what his specific requirements are.

The Realistic DX-150A (Fig. 4-1) is sold by Allied Radio Shack. It covers a wide range (535 kHz to 30 MHz) in only four bands. However, a bandspread control is included so the coverage can be "spread out" on any one band. The bandspread dial is calibrated for the amateur bands, but you can use it on any band such as the short-wave broadcast bands. The receiver is fully solid-state, containing some 17 transistors and 13 diodes and is of a conventional superheterodyne design with a 455 kHz IF. It can receive SSB as well as CW signals and offers other useful features such as an ANL (automatic noise limiter) and "S" meter. It even has an AVC time constant switch which is unusual on receivers of this type. The switch allows the selection of a fast AVC for SSB or CW and a slow AVC action for AM.

Fig. 4-2. Specifically designed for the amateur bands, the Lafayette HA800 tunes from 6 through 80 meters.

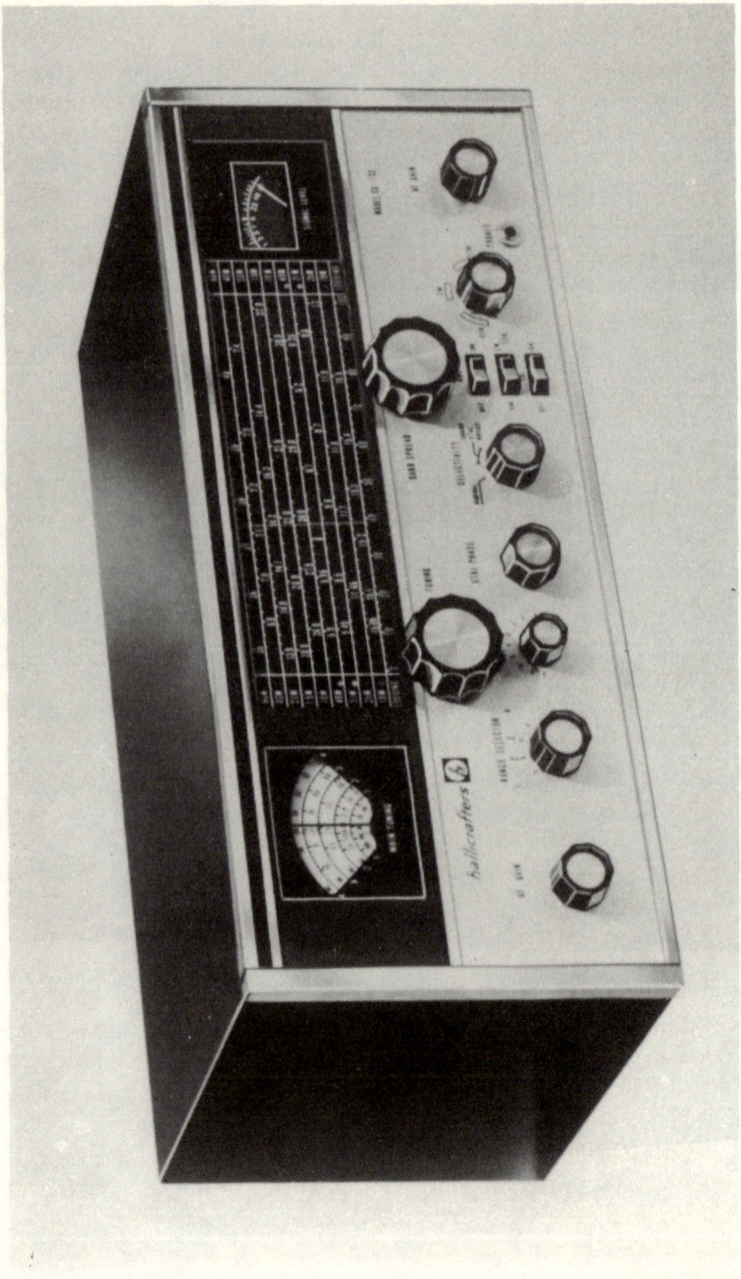

Fig. 4-3. The Hallicrafters SX133 covers the entire short-wave range continuously but with specific calibration for the amateur and international broadcast bands.

Fig. 4-2 shows the Lafayette HA-800 receiver. This is an amateur band only receiver with wide coverage from the 3.5-MHz amateur band (80 meters) to the 50-MHz amateur band. A double conversion superheterodyne system is used with IF frequencies of 2,608 and 455 kHz. Two mechanical filters improve selectivity (automatically switched when the function switch is on AM or SSB). The unit is completely solid-state, including several field-effect transistors (FETs). It can be operated either from 117 volts AC or from a 12-volt DC battery.

The Hallicrafters SX-133 (Fig. 4-3) is an advanced type of receiver covering 1.7 MHz to 31.5 MHz in four ranges. Detectors are built in for SSB, CW and AM. The bandspread arrangement is a little different in that the main tuning scale is in the smaller window on the left. The bandspread tuning, which is located across most of the receiver, includes calibrated scales for the 80, 40, 20, 15 and 10 meter amateur bands, 11 meter CB and the 49, 31, 25 and 19 meter international broadcast bands. The selectivity control is not tied to any other control and is separately available on the front panel. A crystal filter is used for increased selectivity on CW or SSB signals.

The Hammarlund HQ-215 in Fig. 4-4 is capable of receiving any frequency within the range of 3.4 to 30 MHz in 24 segments of 200 kHz each; that is, the receiver is not continuously tunable over the 3.4 to 30 MHz range, but you can select specific portions of this total range for coverage in 200-kHz increments. For instance, you might select 3.8 to 4.0 MHz in the amateur 80 meter band, 5.5 to 5.7 in the aeronautical band, 8.0 to 8.2 for maritime communications, etc. This selection process is made possible by double conversion where one of the mixer oscillators is crystal controlled. The advantage of this system is that each 200-kHz segment can be covered by a very accurately calibrated form of bandspread tuning which allows a frequency readout accuracy of 1 kHz or less. The stability of the receiver is also greatly improved.

For the hobbyist who has decided that his receiving interests lie in specific portions of the high-frequency spectrum, this type of receiver offers an optimum type of performance second only to very expensive commercial designs. The HQ-215 is fully transistorized and has IF frequencies of 3,055 and 455 kHz. A multiplier is used in the 455-kHz IF as a help to the

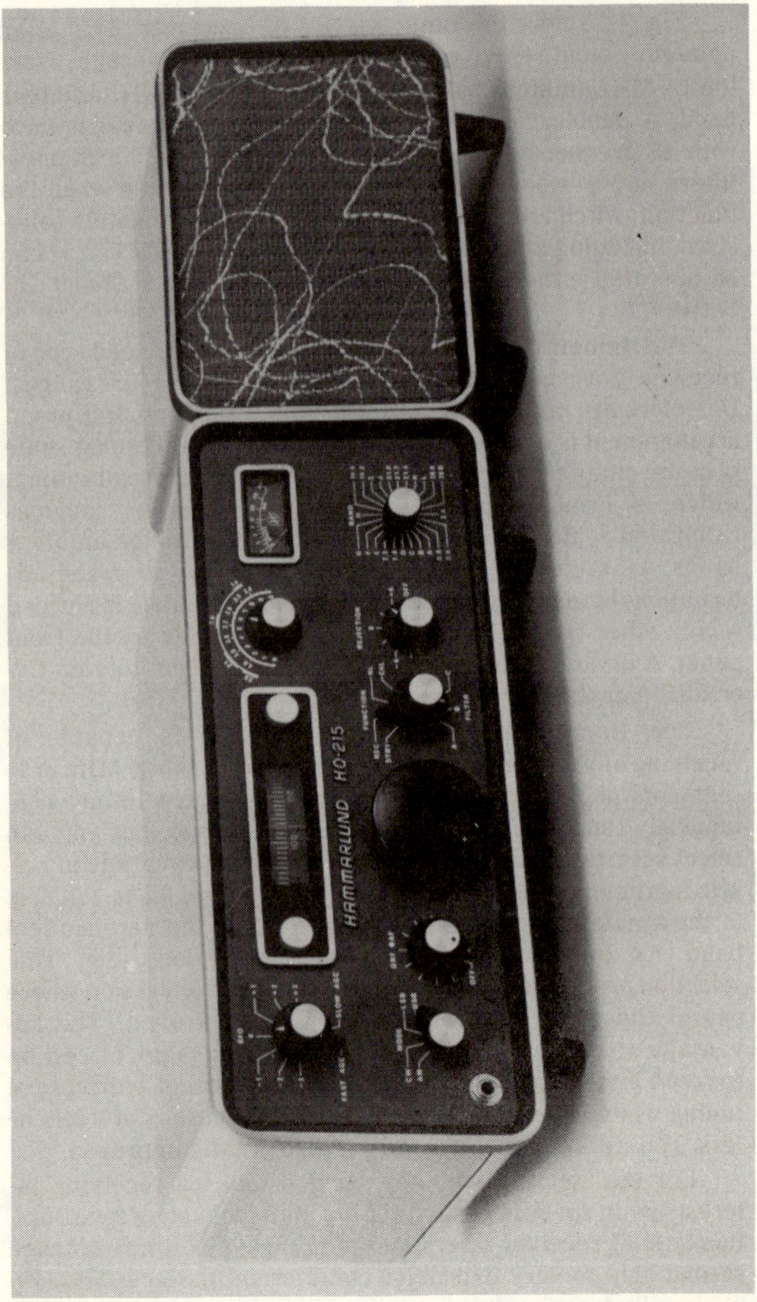

Fig. 4-4. The Hammarlund HQ215 is designed to tune to 24 selected portions of the 3.4- to 30-MHz range.

mechanical filters (6 kHz for AM, 2.1 kHz for SSB and 0.5 kHz for CW) which provide the main selectivity. The sensitivity is less than ½ microvolt for a 10 db signal-to-noise ratio on SSB or CW.

SINGLE-BAND AND FIXED-FREQUENCY RECEIVERS

Single-band and fixed-frequency receivers generally operate in the VHF bands, although there are a number of exceptions. For instance, many marine receivers are fixed-frequency types but part of a transmitter-receiver combination operating in the 2-3 MHz marine band. Special single-frequency receivers are available in the short-wave range, but their application is usually restricted to commercial interests. One exception to this might be the WWV-only type of receiver which has crystal-controlled positions on 5, 10, 15, 20 and 25 MHz. For the very advanced hobbyist who wishes to quickly check propagation conditions in various portions of the short-wave range, such a receiver may be of interest.

The Squire-Sanders "FM-Alert" is, however, more typical of the single-band receivers in use (Fig. 4-5). It covers only the low-VHF public safety band with continuous tuning of the entire band and quite accurate frequency calibration. It also has the interesting feature of being able to switch between continuous variable tuning or to any one of two crystal controlled positions. Thus, you can equip the crystal-controlled positions for reception of known local public safety frequencies, while the continuous tuning feature can be used to explore the band for other reception possibilities. Other controls on the receiver include volume and squelch adjustments. The squelch control sets a threshold level, silencing the audio output until the received signal is strong enough to exceed the set level. Then the audio channel functions normally. This feature is particularly useful when using the crystal-controlled fixed-frequency channels, since you know local signals will always be fairly strong.

In contrast to the previous receiver, which is designed as a fixed base station receiver, the Allied VHF Monitor (Fig. 4-6) is a portable receiver, very similar in appearance to an ordinary transistor radio. However, the receiver is really quite different, since it is designed to cover only the low VHF band (27 to 50 MHz) or the high VHF band (146 to 175 MHz). The units are completely solid-state, of course, with built-in

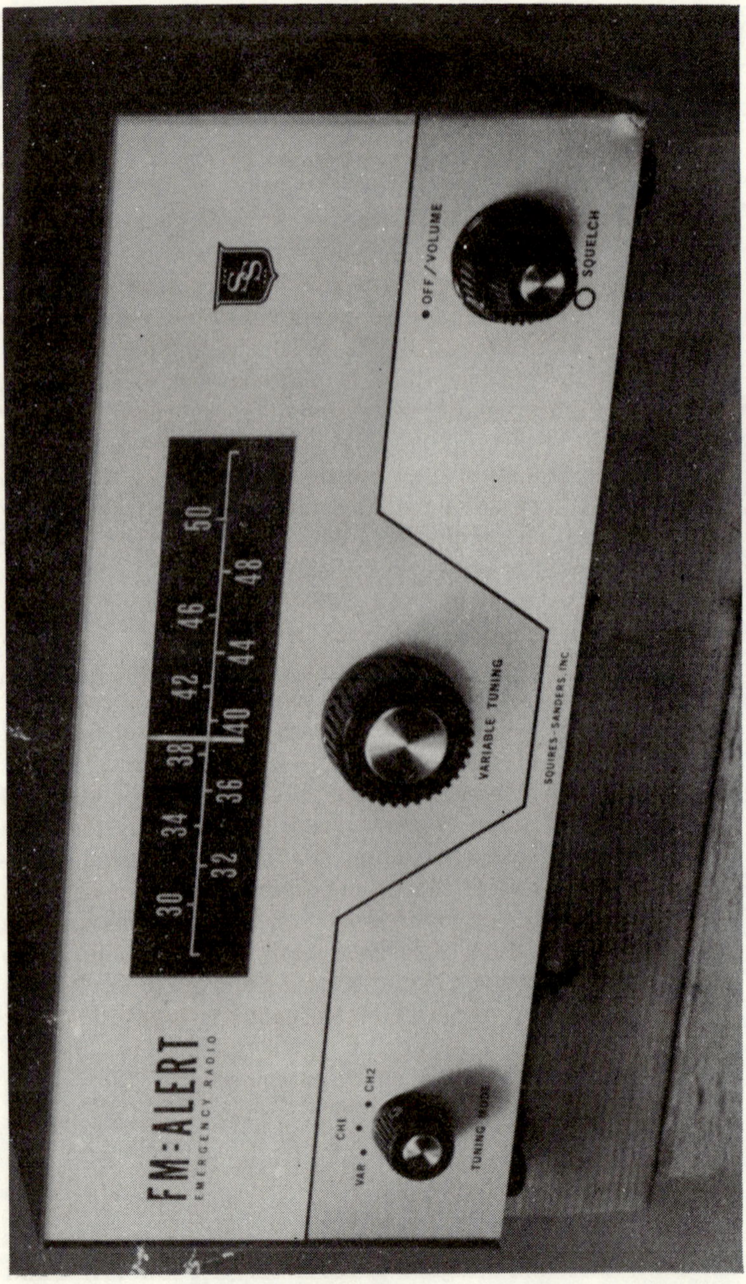

Fig. 4-5. Squire-Sanders FM Alert is a highly sensitive receiver designed for base station reception of a single VHF public service band.

Fig. 4-6. Allied Model A-2587 VHF FM receiver is designed for reception of a public service band.

Fig. 4-7. The Telsat 150 combines the functions of a CB transceiver (with 23-channel crystal control) and a tunable high VHF receiver.

speaker and whip antennas. Some features are naturally lacking in such units, but this is completely understandable for such portable units. The frequency readout can only be approximate and such features as squelch or crystal control are not included. Nonetheless, for portable monitoring of local VHF frequencies, such units are hard to beat.

RECEIVERS WITH EXTRA FEATURES

The variety of receivers with extra features is almost endless. In the short-wave range, receivers are available with built-in cassette tape recorders, for instance. In the VHF range, extra features often have to do with increased tuning versatility or band coverage. For instance, the Lafayette Telsat 50 or 150 are basically CB band transceivers, but they also include completely separate tuning of either the 30 to 50 MHz or 150 to 174 MHz bands (Fig. 4-7). Not only do they tune each VHF band, but crystal-controlled positions are provided in each band for two spot frequencies, which you would normally use for local public safety channels.

For the more serious monitoring operator a particularly popular type VHF receiver is a "scanner." Many such receivers are available, but the Electra Bearcat, shown in Fig. 4-8, is one of the original and most versatile. Basically, the receiver is a fixed-frequency crystal-controlled unit that switches automatically and continuously from channel to channel until a signal is received. The receiver then "locks" on that channel as long as a transmission is taking place. Afterwards, it continues to automatically switch from channel to channel. The Bearcat Series receivers are all advanced double conversion units with IFs of 10.8 and 1.65 MHz. They use FETs in the front-end and are available in models to cover the 30 to 50 MHz, 150 to 174 MHz, and 450 and 470 MHz bands. Such receivers are extremely fine units for someone who knows which channels in each band he desires to monitor; thus, the expense of having separate monitor receivers tuned to each channel is eliminated.

For someone who is perhaps not quite so sure which channels are receivable locally, a dual band VHF receiver such as the Allied Monitor III (Fig. 4-9) has a number of advantages. The unit is completely solid-state and tunes both the high and low VHF public service bands. The unit is complete in itself with a built-in speaker and power supply. Basically,

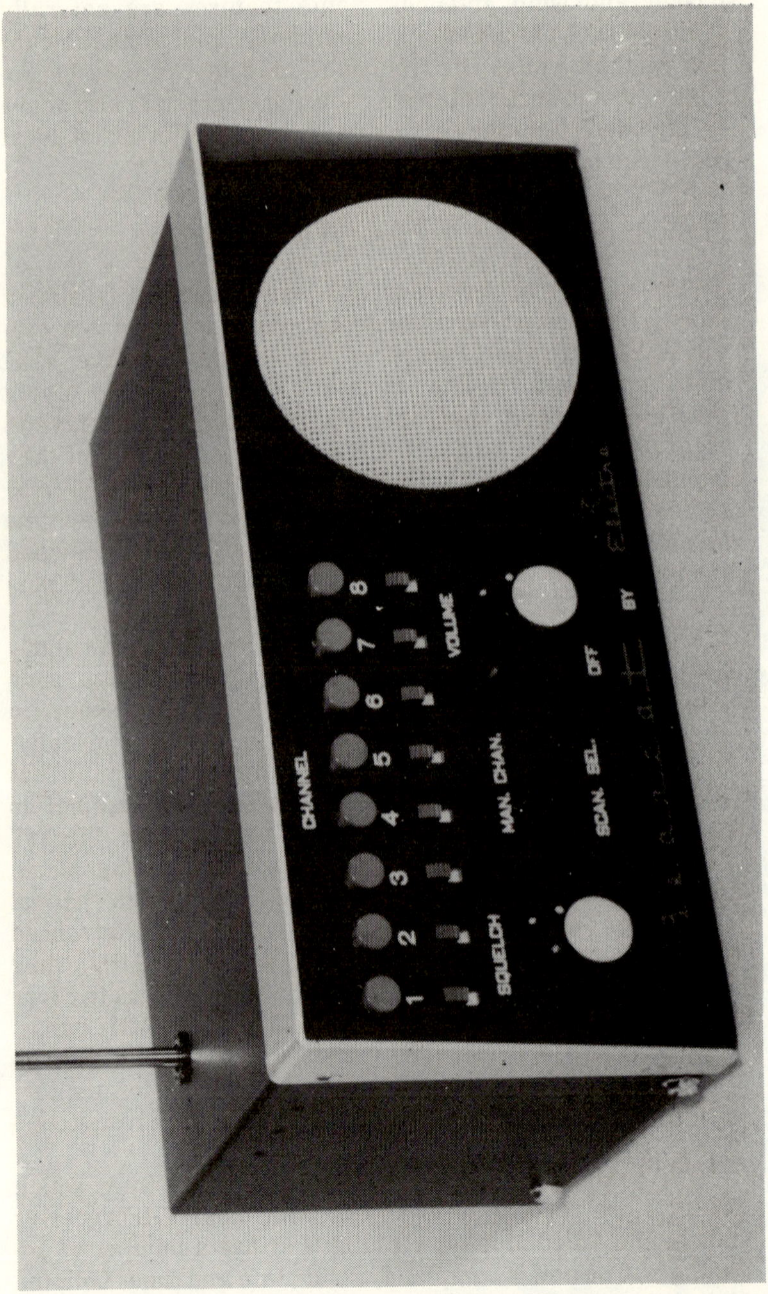

Fig. 4-8. Bearcat scanning receiver offers eight fixed frequency channels.

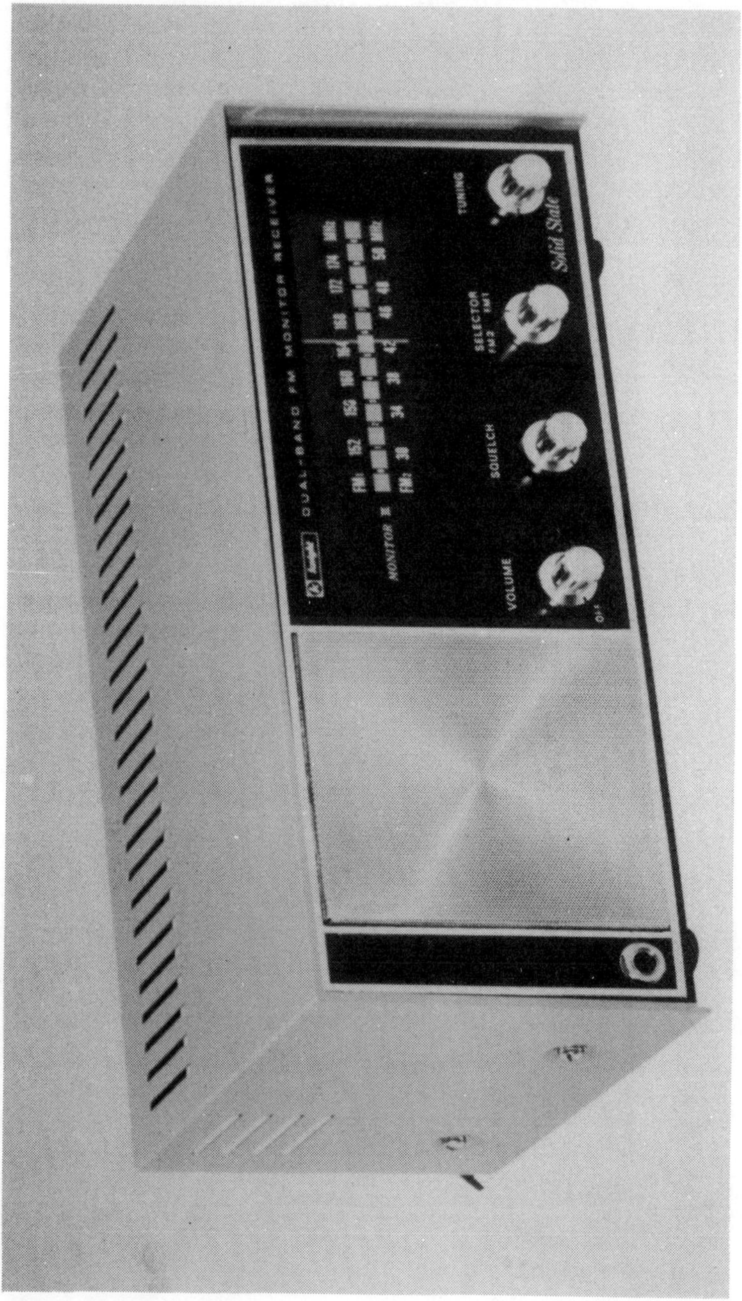

Fig. 4-9. Allied Monitor III provides selectable reception of either the low or high VHF public service bands.

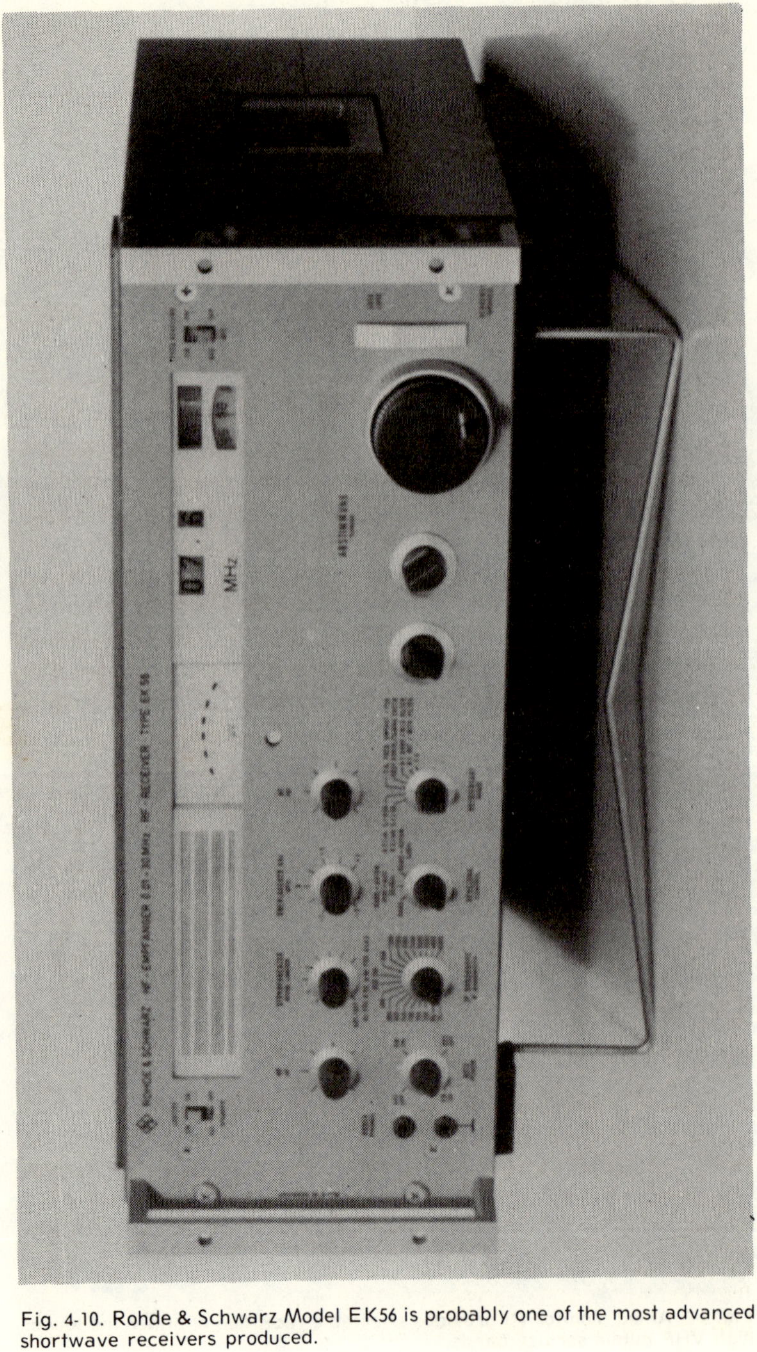

Fig. 4-10. Rohde & Schwarz Model EK56 is probably one of the most advanced shortwave receivers produced.

all that is needed is a simple antenna. The unit comes in kit form at a very reasonable price. However, even beginners need not be scared because of the "kit" aspect. Almost all of the components are already soldered to circuit boards and what is left is mostly just a final assembly procedure of the circuit boards with the chassis and enclosure.

SPECIALIZED AND COMMERCIAL RECEIVERS

Disregarding price tags for the moment, the state of receiving equipment development cannot really be appreciated without becoming familiar with a number of commercial designs. Some were developed for specialized military purposes and money was not a deterrent to their initial development. Few hobbyists will be able to afford these units but, nonetheless, they are worth considering for their extreme technical superiority.

Fig. 4-10 shows probably the most expensive (about $9,000) short-wave receiver being produced today, the German Rhode and Schwarz Model EK56. It would hardly be possible to describe all the features of this receiver here without taking up excessive space. But, basically, it is a fully transistorized receiver covering the extreme end of the VLF range (10 kHz) continuously through the upper end of the HF range (30 MHz). The entire range (10 kHz to 30 MHz) can be covered in individual bands of either 100 kHz or 1 MHz. Twenty different selectivity positions are available with bandwidths from 75 Hz to 6 kHz. Up to three IF frequencies are used and the receiving sensitivity is rated for CW as an extremely conservative 0.8 microvolt for a 20 db signal-to-noise ratio at a 300-Hz bandwidth. The frequency readout is extremely exact. You can easily read the frequency down to a fraction of 1 kHz even at 30 MHz. For instance, without any strain whatsoever, it is possible to tell that a station is operating on 21.4553 MHz. In view of all these features, this is a professional receiver. You have to know how it was designed and how it operates to really be able to use it to maximum advantage.

Fig. 4-11 shows the Communications Electronics Inc. 521A receiver. It is an advanced commercial design using double conversion (10 MHz and 455 kHz IFs). It continuously tunes from 20 to 70 MHz. AM, CW, FM and SSB can be received within this range. Some 80 transistors, including a number of

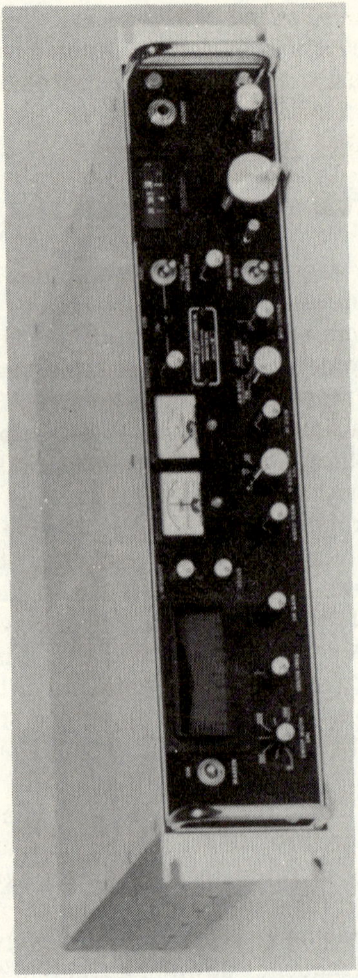

Fig. 4-11. The CEI tunes continuously from 20 to 70 MHz. Received signals are displayed on an oscilloscope.

FETs, are used in the receiver. A built-in panoramic adapter (screen on the left) provides a visual display of the characteristics of the received signal. Modulation percentage, bandwidth, distortion, etc., are all characteristics that can be studied by an experienced operator.

Back to the short-wave range for the receiver in Fig. 4-12. The RF Communications RF 505 is a completely crystal-controlled receiver (via a crystal synthesizer) that provides

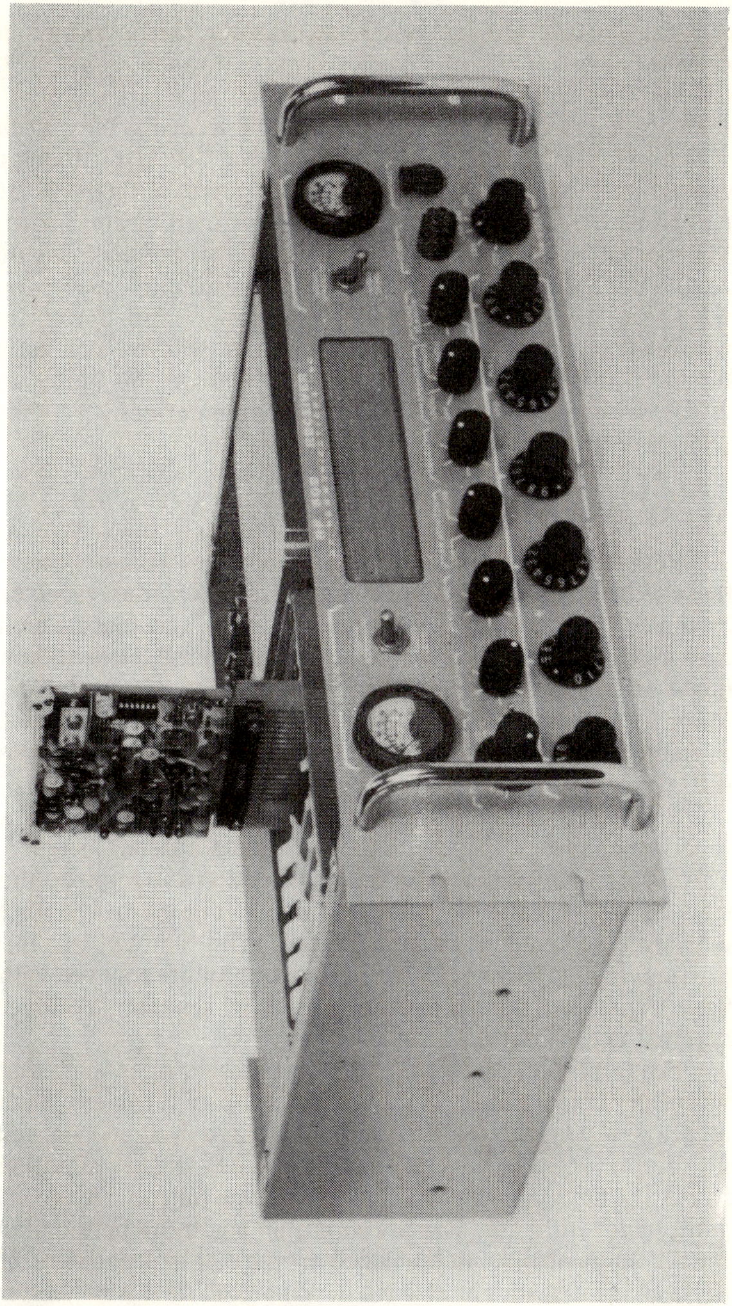

Fig. 4-12. RF Communications receiver, Model RF505, covers a range from 100 kHz to 30 MHz in switch selected steps of 1 kHz.

coverage of the entire 100-kHz to 30-MHz range in switch selected positions of 1 kHz each. All normal modes (SSB, AM, CW) in this range can be received. The unit is completely solid-state with a sensitivity of under 0.8 microvolts for a 10 db signal-to-noise ratio at a CW bandwidth of 500 kHz. Modular construction is used so that any major section of the receiver can be repaired by replacing a simple printed circuit board. The operation of the frequency tuning arrangement should really be appreciated. The operator need only place the frequency knobs on 2.005 kHz, for instance, and the entire receiver including all front-end circuits will be correctly tuned. Within a few seconds, you can change the tuning to 21.005 and again all circuits will be automatically retuned about as fast as you can set the controls.

ACCESSORY ITEMS

Receiver accessories can be divided into two categories: those which improve the performance of an existing receiver by making up for some design compromise and those which extend the ability of the receiver to receive some transmission mode or otherwise allow more versatile use of a receiver. Many hobbyists have been able to judiciously add accessories to build a basic receiver into a versatile, high-performance setup.

Probably the simplest accessory that can be used with a receiver operated above 10 MHz is a good antenna coupler. The coupler will increase the image rejection and improve the signal-to-noise ratio by allowing more optimum coupling between the antenna and receiver. The amount of improvement obtained varies, of course, but amateurs even with very expensive receivers have found a separate receiver antenna coupler helpful for DX reception.

Preamplifiers or preselectors are another accessory that will often improve the performance of simpler receivers. Such an accessory acts basically like the RF stage in a receiver and all the factors that go into making an RF stage evaluation have to be considered when choosing such a unit: noise factor, gain, bandwidth, etc. The preamplifier has to be better than the RF stage already in the receiver in all or at least some of these characteristics or else it will do no good. If it is not better than the RF stage, weak signals will appear louder, perhaps, but not have any more favorable relationship to the noise background. If used with a receiver that has no RF stage, substantial improvement in receiver performance can be achieved.

Preamplifiers and antenna tuners must both be built in shielded enclosures and shielded cable must be used to connect them to a receiver to obtain full benefit. Otherwise, signals may still enter the receiver directly from the antenna or other extraneous sources.

Q multipliers are a popular accessory since only one RF lead (to the mixer tank circuit) is necessary to connect them to a receiver. Their usefulness depends upon the IF bandwidth characteristics of the receiver. Generally, unless the receiver has a crystal filter for CW, a Q multiplier will provide a useful improvement in selectivity. However, for phone reception they may not be so helpful; if the bandwidth is made wide enough for such reception, the skirts will flare out worse than they do with the tuned circuits already in the receiver.

Other selectivity devices such as mechanical, crystal and ceramic filters, notch filters, etc., can be added to a receiver, although they require a slight modification of the receiver circuits. Assuming that the selectivity provided by such a device compares favorably with that of the original receiver circuits, the other factor to be considered is insertion loss. Many devices introduce small losses of less than several db and can be inserted between stages as far ahead in the IF chain as possible. When the loss exceeds several db, generally, an additional IF stage is needed to restore the receiver overall gain to its normal level.

Audio selectivity devices are often packaged separately and used as an accessory following the audio output of the receiver. Frequency peaking, rejection and passband devices are available. Such devices can certainly improve the selectivity of a receiver if the IF passband is broad. However, there are several disadvantages associated with introducing extreme selectivity so late in the signal processing line. Generally, if accessory selectivity is preferable, a tradeoff must be made with the work involved of modifying the IF, possibly adding another IF stage, etc.

Noise elimination devices are in much the same category as selectivity devices. You should first, however, consider the nature of the noise to be suppressed. Simple impulse noise, which has a nuisance effect but does not affect intelligibility seriously, may require only a simple diode clipper in the IF or AF circuits to prevent operator fatigue. The effects of more severe impulse noise may be eliminated only by means of a gating-type circuit.

EXTENDED ACCESSORY ITEMS

A frequency marker unit is perhaps one of the most useful accessories for a receiver that does not have a very linear frequency scale. Such a unit produces marker signals anywhere from every 1 MHz to 100 Hz, depending upon the crystals and circuits used. The most general accessory produces 100-kHz markers and is very useful for checking the calibration points on a general coverage receiver. A frequency divider circuit can be added to provide markers at other intervals; every 10 kHz is very useful for calibration points within an amateur or international broadcast band. In more elaborate form, such units can be used to measure the frequency of transmitting stations by accurately determining the difference in frequency between a very stable marker and the transmitted signal.

Converters can be used to extend the frequency coverage of a receiver. A converter may consist of RF and mixer-oscillator stages or only the latter stages; basically, the front-end of a receiver. The same performance characteristics used to evaluate a preselector must be used to judge a converter. Converters can be either crystal-controlled or tunable. If it is crystal-controlled and a range of frequencies is covered by tuning the receiver (which acts as a tunable IF), care must be taken that signals do not enter the receiver directly by providing proper shielding. Although a converter extends frequency coverage, it cannot compensate for any shortcomings in the receiver it is used with, such as drift and lack of selectivity. The only exception perhaps is that if a receiver drifts excessively on one band a converter might be used to frequency translate that band to another range on the receiver where drift is not a problem.

Multicouplers allow the connection of two or more receivers to one antenna while providing isolation between receivers and additional gain so no power loss occurs in the multiple distribution of the input signal. Generally, such couplers have an RF amplifier stage and a coupling stage for each receiver output. This type of coupler should not be confused with the "couplers" used in FM or TV receivers which contain only passive components and introduce varying degrees of insertion loss. Since the RF stage in the multicoupler becomes the first stage in the signal path, all the considerations applicable to RF stages in general apply and must be considered if the multicoupler is not to nullify the merits of the RF stages in the receivers. To assure that the noise and overload characteristcis of the multicoupler exceed those of a receiver, triodes or silicon transistors in a push-pull amplifier arrangement are frequently used in the RF stages. The isolation characteristics, usually 60 db or more, insure that the oscillator signals which a receiver may couple back to its antenna input do not interfere with another receiver connected to the multicoupler.

Radioteletype reception is often one of the most interesting types possible in the short-wave range. Basically, all that is required for such reception is a good stable receiver, an FSK (frequency shift converter) unit and a teleprinter. Frequency shift converters can be either of the IF or audio type. Such converters vary greatly in their features and range of signal inputs. For the hobbyist, a converter designed for the reception of the standard 850 Hz shift is generally adequate to provide access to an interesting variety of weather and news transmissions.

Selective calling attachments disable the output of a receiver until a coded signal is received which energizes the receiver. A simple form involves the use of a single tone which activates a frequency-tuned reed relay. When a tone of the correct frequency is received, the relay is actuated, turning on the audio output of the receiver as well as locking itself in an "on" position. When more receivers are involved than can be handled by single tones, multiple tones are used and reed relays are chosen so that the correct tones in the correct order (or simultaneously) must be sent to activate the receiver. Still another system transmits pulse groups. The receiver counts the pulses received and is activated only when the correct number of pulses is received.

Audio recorders, connected to the headphones or loudspeaker output, are found very useful by many hobbyists to record received signals. A recording may be used to play back a transmission to insure correct identification of a call sign or other part of a transmission. For voice reception a wide variety of inexpensive battery and AC operated recorders are available.

The panoramic display unit gives a visual presentation of the signals coming through the front-end of a receiver. When used properly, it can be one of the most useful aids for reception purposes in the sense that you can "see" as well as hear a signal. Basically, a panoramic unit is a simple superheterodyne receiver with a cathode-ray tube as the output device. The oscillator is swept across the receiver IF passband at the same time as the sweeping voltage is applied to the CRT horizontal deflection plates. The vertical deflection plates are driven by the detector output. The result is an amplitude-versus-frequency presentation. The unit is connected to the receiver ahead of the main selectivity circuits, usually to the first mixer output circuit. The frequency bandwidth that can be viewed is usually variable and depends upon the receiver's front-end selectivity and the characteristics of the panoramic unit; plus or minus 10 to 100 kHz from the IF center frequency is typical. From the visual presentation, an operator can determine signal modulation characteristics, interference beats, spurious responses (images move in an opposite direction as compared to normal signals as the receiver is tuned, for instance), noise interference and other characteristics once he learns to interpret the visual presentation.

Obtaining and Using Receiving Equipment

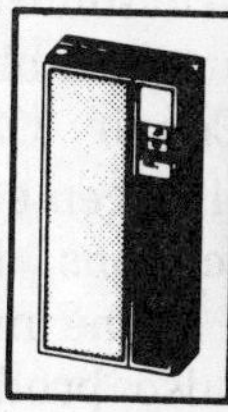

Chapter 5

Few immediately acquire all the equipment or the quality of equipment desired for a receiving setup. Therefore, some planning both as to the use and physical placement of equipment as it is added will eliminate unnecessary continuous rearrangement of the equipment as well as unnecessary drain on the pocketbook. The physical placement of equipment depends a great deal on its use. A CB operator who uses one or more standby receivers to monitor emergency channels certainly doesn't require the same type of operating position as an SWL or amateur operator who might spend several hours tuning and listening to a receiver. Here are some points to look for in purchasing equipment and ways in which receiving equipment might be arranged and used for different applications. Many hobbyists first buy a fairly good receiver, but have no idea of how they will use it and end up being disappointed with the reception they get.

RECEIVER KITS

The variety of new receivers is constantly changing. Therefore, a check-list which will enable you to compare the general features of any receiver on the market should be useful. The best way to keep abreast of new receiver availability in the United States is to send for some of the free large radio supply house catalogs, such as those from Lafayette Radio, Syosset, L.I., New York 11791; Allied Radio, 100 North Western Avenue, Chicago, Illinois 60680, and World Radio Laboratories, Council Bluffs, Iowa. These catalogs will by no means indicate all the receivers on the market, since each company distributes only certain manufacturer's products. However, by perusing the advertisements in radio hobbyist magazines—Popular Electronics, Electronics World, QST, CQ, 73, S9, etc.—you can get a broader view of the models available and send for brochures on models of interest. Kit models are listed in some of the above catalogs and magazines and others are available from Heathkit, Benton Harbor, Michigan, and Conair, 3939 Wisconsin Ave., Washington D.C.

Commercial receivers are not advertised generally in hobbyist publications because of the limited market. However, most of the manufacturers of extensive hobbyist receiver lines also produce various commercial receivers which may be of interest to the hobbyist. Generally, the commercial versions do not provide superior electrical performance but rather superior mechanical performance, since they are designed for continuous 24-hour operation, for instance. For those interested in receiver types available overseas, try to obtain a copy of the British magazine "Wireless World."

The question of kit versus ready-built equipment has several sides to it. Generally, you can obtain a bit more performance per dollar from a kit, although the margin is a narrow one when Japanese imported receivers are considered. United States kits (assembled) do seem to have a better resale value, however, than imported ones. Sometimes, it appears that a kit offers an advantage because you learn something about electronic circuits as you build it. Whether you really learn anything about it depends solely on your patience. You can follow instructions like a robot and simply put the wires between designated points without any idea of what circuit is being wired. On the other hand, by trying to understand the function of each component as it is wired into a circuit, you can learn a great deal about component and circuit functions. Most kit advertisements give an estimate of how many hours it takes to assemble an item. Generally, the estimate is based on a steady effort with no allowance for the time needed to understand the circuitry. If you analyze the circuit while you build, the total time can easily be two to three times the manufacturer's estimate. This total time should be translated by the hobbyist into how many actual evenings of work are required. If you don't have the patience to wait so long to complete the receiver, a kit should be forgotten.

Although almost all advertised kits are simple to construct if sufficient time is taken, there is probably nothing more frustrating than trying to chase the "bugs" out of a too hastily constructed unit. So, if the kit route is taken, by all means be prepared to take your time twice over. Following are some of the main items that should be considered in comparing receivers.

GENERAL ELECTRICAL FEATURES

The simplest and most obvious electrical feature of any receiver under consideration is frequency coverage. If you have a specialized interest such as the amateur bands, international broadcast bands or public safety services, you may be interested in a receiver that covers only those particular frequency ranges. Otherwise, you may be interested in as general a frequency coverage as possible. Most new general-coverage receivers tune the entire range from the broadcast band through 30 or 35 MHz. However, some cover only selected portions of the HF range and, of course, there are a great variety of single- or dual-band VHF only receivers.

Mode Coverage. Again, this factor depends on the hobbyist's interest. For the public service bands, narrow-band FM reception is essential, since this is practically the only mode used. For use on the HF bands, reception of AM, CW and SSB may be desired. If you listen only to the international broadcasting bands, the AM reception capability alone may suffice. Radioteletype reception is normally provided by accessory units and only a few very specialized receivers come with this capability built-in.

Performance Capability. This category includes such areas as sensitivity, bandwidth, image rejection, etc. Performance characteristics are determined both by the type of circuit used (single or double superheterodyne, RF stage or mixer as first stage, type of detector, etc.) as well as the type of components used. An elaborate description such as "multi-conversion, high sensitivity, extremely selective" doesn't really say anything about performance; that can be determined only from numerical data of standardized performance tests.

Probably more tradeoffs must be made in the area of performance qualities versus cost than in any other. There is no universal rule to apply in determining the best tradeoff since this depends upon the use of the receiver and the ability to correct deficiencies by accessory equipment. For instance, if you are mainly interested in AM reception, an expensive CW bandwidth filter is a needless feature. If a receiver is satisfactory in most respects except image rejection, the use of a simple antenna coupler as an accessory to improve this condition might be far less expensive a solution than going to a higher priced receiver.

Unfortunately, "Consumer Reports" cover nothing but entertainment receivers. Whether deliberate or not, however, there is just as much use of superlatives and incomplete specifications made when advertising receivers as in any other field of merchandising. For instance, a receiver description might state: "1 microvolt sensitivity with a 3-kHz bandwidth and high average image rejection of -40 db," all of which doesn't say anything really. The 1 microvolt sensitivity must be related to a specific signal-to-noise ratio and bandwidth. The 3-kHz bandwidth says nothing about "skirt" selectivity, only that at some indeterminate point the bandwidth is 3 kHz. The average image rejection by itself doesn't sound too bad for a simple receiver, but the word, average, means it could be only an extremely poor 10 db at one end and a good 80 db in the AM broadcast band. The same holds true for other receiver specifications.

Usually, the receiver operating manual can be purchased separately at a modest price (one to three dollars) by writing to the manufacturer directly. The manual will usually contain a great deal more detailed information about a unit than appears in any advertisement and it should indicate the exact circuitry used. You should also write the manufacturer directly when a specific characteristic—stability, for instance—is not stated and is of importance to the user's intended purposes.

Magazines also provide test reports on receiving equipment from time to time, but sometimes will not highlight any defects too sharply because the periodical cannot jeopardize the manufacturer's advertising account. Therefore, such reports must be read "between the lines," but they still can be of general use. The recommendations of local hobbyists can be useful concerning the features of a receiver, but generally are subjective regarding electrical features since each person's receiving conditions (antenna, siting, noise conditions, etc.) will differ unless the person actually has used good test equipment to measure the receiver's performance.

GENERAL MECHANICAL FEATURES

The mechanical aspect encompasses such items as the tuning mechanism action, placement of controls and frequency readout scales and the structural and electrical quality of the enclosures. The tuning mechanism characteristics, such as kHz per knob revolution, must be described in numerical terms. However, there is also the often used term of "feel" to describe a tuning mechanism, meaning the smoothness of rotation. Being largely a subjective quality, there is no way to judge it other than by actually trying a receiver. The placement and adequacy of the controls and scales can best be judged by visualizing that you are actually using the receiver for the reception of various stations or modes. For instance, on some receivers the main tuning controls are on one side and so high that for table-top use your arm would have to almost be held in the air to tune the receiver—hardly a comfortable position for exploring the amateur or international broadcasting bands. The quality of the enclosure is not too easy to judge, since rarely, except for military applications, are the structural and electrical properties specified. The best the hobbyist can do is to judge the enclosure from appearance, whether or not it is rigid enough for its application (mobile use, for instance) and whether or not it provides adequate shielding by not having any large open areas on any side.

Accessory Provisions. A basic receiver may have no built-in accessories, such as a frequency calibrator or marker, but many do provide for the addition of such an accessory. Other receivers are designed to accommodate even more accessories such as bandwidth filters and converters to extend the frequency coverage of the receiver. If you plan to increase the versatility of a receiver, such "plug-in" accessory features can save a great deal of work even if you intend to home-build an accessory or use one made by a company other than the receiver manufacturer.

Compatibility. Not many people have purchased diesel-powered automobiles thinking they were gasoline types, but it has happened. People also buy receivers whose power requirements are not compatible with the source available, be it 110 volts AC or DC, 180 or 220 volts AC, 6 or 12v DC, etc. Various forms of converters are available, of course, to make any receiver work with almost any source voltage. Severe line voltage fluctuations may also require the use of some form of regulator to maintain proper receiver gain and stability. Physical size and weight is not often a problem but it does have to be checked for mobile applications where physical interference and support factors must be considered. Material usage is important only if the receiver is to be subjected to unusual environmental conditions, such as salt or tropical conditions, where inert or protected metals must be used.

PRICE TIPS

Another area to be checked, especially if you are buying a new receiver as a first receiver, is the general resale price of receivers in the same price range from different manufacturers. Generally, a good resale price augurs for a quality product.

When actually purchasing a factory-built receiver, you should shop around for the best prices if paying cash. Radio periodical advertising often calls attention to small dealers who specialize in mail order sales and who will give discounts; usually, though, "service" ends at this point. A postcard survey of these dealers can often result in profitable savings when higher-priced equipment is being purchased, but such savings should be weighed against the convenience and service supplied by a regular franchised dealer. Except on close-out sales, discounts on kits are not generally available.

SURPLUS AND USED RECEIVERS

Surplus receivers are generally World War II gear although a number of military designs constructed in later years often appear on the market in limited quantities. Used receivers are commercially manufactured types, either designed for hobbyist or commercial use, that are offered for resale. Whether "surplus" or "used," they still have to be evaluated as previously described regardless of the seemingly bargain price. In addition, a few special considerations are in order.

Surplus military receivers, for instance, were constructed for specific installations and are not necessarily usable "as is" even though in new condition. Perhaps the simplest consideration is the power source requirement, which may vary widely from normal 60 Hz 110v AC to 110v AC 400 Hz or 24v DC. Separate "tuning heads" are necessary with some receivers because they were designed primarily for remote control. Others will require circuit modifications for the same reason or because they were designed for a particular mode only or because they were wired to interconnect to a companion transmitter.

Exact specifications are rarely available for surplus equipment beyond the generalities of frequency coverage and main circuit features. The tubes and parts used in older surplus equipment may no longer be readily available should they require replacement.

These points are not meant to imply that surplus equipment should be forgotten; such equipment may well present excellent values dollar-wise. However, before purchasing such equipment the tyro hobbyist should examine one of the surplus conversion manuals sold and determine if the work which has to be done to make such equipment fully operable is within his capability. You should also realize that the resale value of all but a few surplus receivers is relatively poor. Many dealers will not take surplus equipment as trade-ins toward the purchase of new equipment.

Purchasing a used receiver involves much the same considerations as buying a new receiver, with the exception that you must determine how closely the present condition of the receiver, both electrically and mechanically, approaches its original condition. An operating manual may come with the receiver or should be obtained from the manufacturer in order to determine the original condition of the receiver. For older receivers such manuals may not be readily available and the purchaser must rely upon his evaluation of the receiver as it exists. Many used equipment dealers claim to recondition their equipment and most probably make an honest effort to do so. The physical condition of equipment (mars and scratches) may be obvious; however, no used equipment comes with any summary of its electrical performance statistics, so who knows exactly? The situation becomes particularly complicated if the original owner made any modifications to the receiver circuitry.

Probably the best protection when buying a used receiver is a clear guarantee as to under which conditions the receiver may be returned for a refund if found unsatisfactory. Most dealers allow the return of used equipment but only provide a credit towards purchase of another used or new equipment of an equal or higher price rather than a cash refund. Discounts may be given on a cash deal for used equipment. Therefore, you should certainly check the price of used equipment with as many dealers as possible to obtain an average price. In spite of the goodwill of any number of honest dealers, the phrase "buyer beware" could never be more applicable.

USING THE RECEIVER PROPERLY

One of the most useful things you can have for any receiver is a copy of the original operating manual. However, such manuals are not always available and some are not too detailed, so a discussion of some details on the usage of the controls on a typical receiver is appropriate. The optimum setting for the controls on any receiver depends upon reception conditions and the characteristics of specific receivers, so some experimentation has to be done once reasonably good reception is obtained. The following points regarding specific controls should be noted:

Calibrator. If a built-in calibrator or frequency marker (usually 100 kHz) is available, it can be used to check the main tuning to the nearest 100-kHz point by turning the BFO on (centered, if variable) and tuning for zero beat. If the frequency markings on the main tuning scale are not very spread out, it is a good idea to be sure that the main tuning is set to the correct 100-kHz point for bandspread purposes by checking the scale calibration by receiving a station of known frequency. Once the main tuning is set, the bandspread tuning can be calibrated by noting where the 100-kHz calibration points occur on the bandspread scale. With many receivers the range covered by the bandspread changes considerably throughout the range of the receiver. Therefore, calibration of the bandspread scale must be redone everytime the main dial position is changed.

AGC (AVC) Selection. Some receivers provide various discharge time options for the AGC circuit (usually labeled slow, medium and fast). If the AGC is left turned on for SSB and CW reception, either a slow or medium AGC time constant will usually prove best. Usually, it is possible with such AGC circuits to run the RF gain control at maximum on all modes. For AM reception a slow AGC time constant is usually used, but under conditions of rapid fading, either medium or slow time constants may prove useful.

Selectivity Devices or Bandwidth. Slot filters, Q multipliers and phasing controls for crystal filters should be left off until needed or else they will confuse the tuning. Once a station is tuned in as well as possible, and interference still is a problem, the selectivity device should be tuned slowing for the best tradeoff between interference rejection and signal intelligibility. Generally, retuning the main tuning control and selectivity device simultaneously will not help. With a Q multiplier the receiver noise will decrease as selectivity is increased due to the narrowed bandwidth. It does not indicate any decrease in receiver sensitivity. Some few receivers use the Q multiplier as a BFO by allowing it to go into oscillation. Such usage eliminates its usefulness as a selectivity device when it functions as a BFO for CW and SSB reception.

The BFO control on some receivers is actually a switch used to change the crystals in the BFO circuit for USB, LSB or CW reception. When the BFO is crystal-controlled, you have no option but to use the main tuning control to achieve the desired tone for CW reception or the most intelligible SSB reception. On other receivers, the BFO is variable plus or minus 3 to 6 kHz from the IF frequency. In this case the main tuning control should always be adjusted for the loudest signal about 1 or 1.5 kHz off the center frequency. The side of the center frequency used depends upon whether LSB or USB signals are being received.

Squelch controls are normally found only on VHF and mobile receivers. They simply act to cut off the receiver output until a certain input level is reached. Normally, the control is adjusted so that the receiver output just cuts off with the normal no-signal noise background. Adjusting the control too far will mean that the receiver will remain cut off until a very strong signal is received and will not come on for weak signals.

Mode or Frequency Control. Some receivers offer both variable tuning or crystal-controlled operation and there is a control so labeled. In the tunable position, of course, the receiver's variable oscillator is used to allow continuous frequency tuning. The crystal-controlled position provides reception of one specific frequency per crystal switch position at usually much greater stability than provided by the continuous tuning. The frequency of the actual crystal used for reception of a specific frequency depends upon the circuit design of the receiver. Many times, the circuit design is such that the crystal must be chosen specifically to be above or below the frequency input to a converter stage for best results. Information on selecting the proper crystal will usually be contained in the receiver operating manual.

Function switches may be labeled send-receive or standby-operate. In either case when the receiver is used alone, the switch is normally in the receive position. When used with a transmitter, the switch would be in the send position, which closes relay contacts that connect together the receive terminals on the rear of the receiver (the terminals parallel the receive position switch contacts) when the transmitter is inoperative.

Tuning or level meters are useful to determine when the receiver is tuned for the loudest signal. Depending on the receiver's circuit, they may be operative on AM only or on CW and SSB as well as FM (in the latter case it would be labeled a discriminator meter). Under conditions of heavy interference the meter readings are usually too erratic to be useful and you must depend strictly on aural tuning. The scale calibration on such meters will be accurate only under specific conditions, usually as outlined in the receiver's operating manual, and should not be taken as absolute readings. Depending upon the meter characteristics and the circuit, a meter may appear to read low on relatively strong signals or bounce very lively on even weak signals. Neither indication necessarily indicates whether the receiver is sensitive or not.

The controls on many receivers combine some of the functions mentioned here. The control for mode selection, for instance, might simultaneously control the AGC, BFO and IF bandwidth circuits. The presence of fewer controls certainly streamlines appearance and for various applications makes faster frequency tuning possible. However, for the hobbyist, a receiver with a switch for each and every circuit function is not necessary, but sufficient controls to permit individual adjustment of all major circuit functions is certainly desirable.

EQUIPMENT ARRANGEMENTS

The electrical interconnection of the components of a receiving system certainly determines the degree of technical performance of the equipment involved. However, the physical placement of the equipment determines the convenience with which the equipment may be used and, in fact, the desire of an operator to properly use the equipment. Optimum use of equipment obviously requires the proper combination of the two preceding conditions.

When two or more receivers are used in a simple setup, they may each be connected to a separate antenna and function independently. If only one antenna is available, the receiver antenna terminals (none connected to ground) should be connected in series to the single antenna and not in parallel. If the inputs are connected in parallel and the receivers tuned to different frequencies, one receiver will tend to short-circuit the other's input. If the receivers do not have two terminal inputs independent of a ground connection (only inputs for unbalanced transmission lines such as coaxial cable), there is no choice but to connect the inputs in parallel or to use an antenna preamplifier with two outputs for separate receivers. The use of a pre-amplifier not only eliminates the slight loss of signal voltage incurred by connecting multiple receivers directly to one antenna, but also prevents the spurious signals from each receiver feeding into the other antenna input. If there is no problem with spurious signals from either receiver (checked by connecting their antenna inputs together and tuning each receiver through its operating range), an even simpler resistance dividing network can be used for multiple receiver outputs as in Fig. 5-1.

If two receivers are used for some form of diversity reception—either space diversity where two separate antennas are used or frequency diversity where the same transmission is received as it is broadcast simultaneously on two different frequencies—the audio outputs of the two receivers must be combined. This can be done simply by connecting both audio outputs in series or parallel to a common loudspeaker. Some audio loss will occur but most receivers can supply enough audio output so that the loss is not objectionable. To achieve a true diversity effect, the output of the receiver which at any instant receives the weaker signal should be suppressed. One simple way to do this is to connect the AVC lines of the two receivers together. If the AVC characteristics of both receivers are identical, the method will work fairly well.

As more accessories are added, it may be very desirable to add some "patch" panels for the receiver input and output

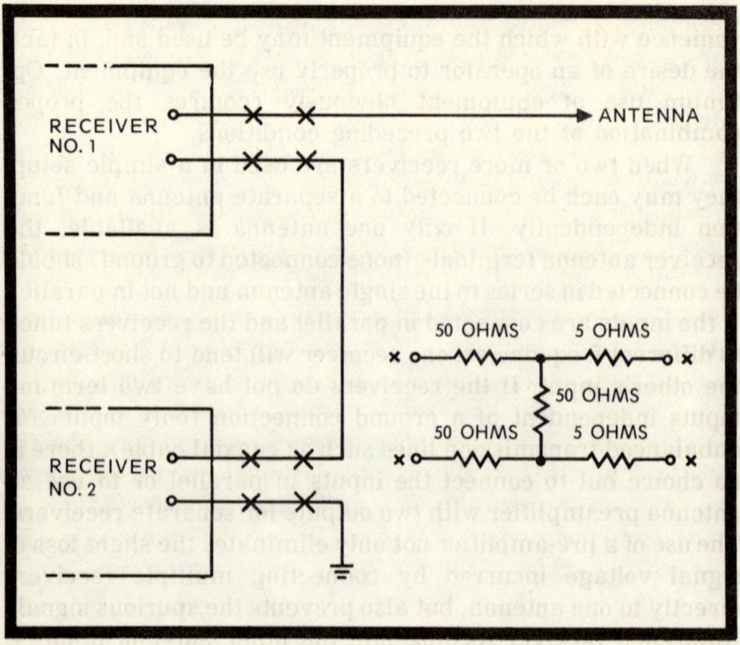

Fig. 5-1. If directly connecting one antenna to two receivers causes too much interaction, insert the resistor pad at points marked X. Resistor values need not be exact.

functions. Such panels simply "extend" the antenna terminals, headphone terminals, accessory terminals, etc., of one or more receivers to a common area where accessories can be easily switched in or out of any circuit. Tests can be easily done to determine which unit is not working properly in case of a failure or to measure equipment performance levels. New accessories or circuits can be tried out very easily without the usual need of continually disconnecting wires from the receivers. If accessories such as panoramic adapters, radioteletype converters or IF filters are used, which require access to IF circuits, another patch panel for IF switching might be added. Care must be observed, however, in doing this. Some accessories, such as Q multipliers, will not work properly if the lead to the IF stage is too long.

The patch panels themselves can be constructed with switches, if desired, but you can rarely immediately define all the switch positions desired. Generally, a system of plugs and jacks with short jumper cables to make interconnections is both less expensive and more versatile. The plugs and jacks for the audio switching can be the standard PL-55 types found on headphones, for instance. Or, RCA type phono plugs and jacks can be used for both audio and antenna switching; however, those used for antenna switching should be the type with nylon insulation designed for RF use. Still another possibility, if a compact panel is constructed out of some insulating material, is to use simple screw-type binding posts instead of plugs and jacks.

One precaution to observe when switching audio functions, particularly from tube type receivers, is to insure that unused audio outputs from a receiver or recorder are terminated in a resistance of 50 ohms or less. Such a termination prevents insulation breakdown in the output transformer when amplified signals produce high open-circuit voltages in the unloaded transformer. You should also provide a reasonable number of extra patch panel positions to allow for the addition of future accessories.

Most hobbyist receiving installations develop rather haphazardly and the result is often a "rat's nest" of units and wiring. This can be avoided by planning the operating position in advance. Also, a more comfortable position for the operator will result. Of course, the type of human engineering that goes into planning equipment installations for space craft or military installations where an operator must spend hours or days at a position is hardly necessary. But a little thought as to placing receivers where dials and controls can be easily read, where switches are in convenient positions, proper lighting, etc., will definitely pay dividends.

You can purchase operating benches made of wood, which come in unpainted form much like furniture, or commercial type sloping-front panel metal enclosures which can be grouped in a semi-circular form for very elaborate installations. However, most hobbyists settle for considerably less, usually placing the equipment on top of a desk or work bench. However, even in such a situation, the construction of a simple equipment shelf can be very useful. Such a shelf should elevate the equipment 5 to 6 inches with a slight backward slope. It allows the receiving equipment to be elevated enough so the dials are more nearly at eye level when the operator is seated. Power outlet strips and a lamp underneath the shelf are handy convenience features. Small accessories can be placed underneath the shelf and easily wired into the rear apron connections on the equipment on top of the shelf. In a more expanded installation, space beneath the shelf could be used for audio and antenna patch panels.

Outlets along the rear underside of the shelf can be individual convenience outlets, a strip outlet or even a commercial multiple outlet which includes a fuse and line switch so power to all units can be turned off simultaneously. If only enough illumination is desired to read a schedule or use a log book, a small 4- to 10-watt bulb in an automobile interior light fixture mounted on the front edge of the shelf is sufficient. If lighting for prolonged reading is desired, two 15-watt T-8 fluorescent lamps or a single 20-watt unit can be used.

If you have more equipment than can be mounted on one table, an L-shaped arrangement is frequently used. Test equipment and other less frequently used equipment can even be mounted beneath the table on another separate shelf. The interconnecting wires which run behind the shelves between units should be bound together with plastic cable straps rather than electrical tape which is very inconvenient to remove for cable additions.

Rather than a shelf type of arrangement, some operators prefer sloping the front of the equipment about 30 degrees or so from the vertical, while the rear of the equipment rests directly on the table. The front is elevated either by a fixture or by just using extra long rubber feet in front. Although far better than having the equipment rest flat on the table, it does not provide any convenient mounting space for accessories as is the case with a shelf. Accessories must be mounted by the side of the receiver or on top where they are in the path of the heat generated by the receiver.

Some attention should also be paid to the room itself if an installation grows so elaborate that many loudspeakers are operating at once. A room full of hard-surfaced materials will produce sound reflections that soon become tiring. The use of cloth drapes or commercial sound proofing panels will quiet the room.

QSL CARDS AND GIVE-AWAYS

The collection of QSL or acknowledgement cards can become a very interesting aspect of using a receiving setup.

You can specialize in collecting cards from certain areas of the world or countries, from maritime or aeronautical stations, amateur stations, etc. Many DX clubs sponsor contests and-or award certificates for collecting QSL cards. The business of QSL card collection can possibly get out of hand, however. Some magazine articles have gone so far as to suggest soliciting cards from the local police department by listening to their transmissions, telephone company and other business-band stations, etc. While there is nothing wrong with such solicitation, the chances of obtaining many replies in card form is not good. Such stations normally do not have QSL cards printed for distribution. Some hobbyists who insist upon such acknowledgment send the station a blank QSL (with return envelope) which the station can fill in and return. The same practice will also generally produce better returns when sending cards to domestic or foreign maritime, aeronautical, time stations and similar stations who would not normally have a stock QSL.

The best QSL returns will be obtained by sending reception reports to the types of stations who are eager to receive such reports, mainly amateur and international broadcasting stations, and by making the reports meaningful. The eagerness of many international broadcasting stations to receive reception reports is attested to by the fact that they have contests for reception report submittals, offer small giveaways for reports, such as pennants, etc.

There are many reasons why such stations are eager to receive reports. They can provide valuable information about program popularity, audience size, the effect of technical improvements, propagation studies, etc. Various schemes are tried by these stations in order to evaluate broadcasting effectiveness. For example, some announce different mailing addresses on different frequencies or on different programs or on transmissions directed to different geographical areas or to different language groups. Nonetheless, for the QSL collector, they offer some very attractive cards.

A reception report will more easily secure a QSL return if it contains some information of value to the recipient (see Fig. 5-2), whether it is a broadcasting station or not. A reception report that simply says, "your station received on 20 January at 1700 GMT," has limited value. It still conveys some information of value to a broadcasting station when combined

MEMBER'S RECEPTION REPORT

Official Publication "The Australian DX News"

AUSTRALIAN RADIO DX CLUB

Report No.3 .... Date Prepared 1/2/71. .... Membership No. ....

Radio TRIO 9R59 DS .... Name Chris Hambly.

Address 47 Chessell St.
Box Hill Nth.
Victoria. Australia.

Dear Sirs,

I am taking this opportunity of submitting this report of reception of your station Radio Liberty which I monitored on a frequency of 15125 kHz, in the 19 meter band band, from 1100 hours until 1115 hours GMT on the ....

I have listed below details of program items heard, and I would be pleased to receive a verification card, or letter, from you confirming this report.

Date (GMT) 1/2/71		S	I	N	P	O
Time (GMT)	Details of program heard News Commentary	4	3	4	2	3
1100	Sign on with Czech & Slovac Lang with the news severe interfrence from Radio Berlin International on same frequency.m					

Receiver & antenna TRIO 9R59 DS with long wire.

Remarks ....

I trust that this report will be of value to your technical staff, and I hope that you will have continued successes with your broadcasts in the future. Looking forward to your reply.

I remain,
Yours faithfully,

Chris Hambly.

Fig. 5-2. A very neat and practical reception report form. It is printed on thin paper which folds together to form an air mail letter.

with a knowledge of the sender's location, but is of far less interest to a radio amateur. A reception report that really contains more specific information will generally follow the form of the SINPO report, which is explained in Fig. 5-3. Detailed reception information combined with a note about the receiver and antenna used by the listener will really allow a station to evaluate its effectiveness. Information about the frequency of a station even if only approximate is valuable to broadcasters since they may have several transmitters operating simultaneously in one band. An additional note in the reception report about either the technical or program peculiarities or appeal is also a good idea as it concerns modulation quality, distortion, etc.

Sending a reception report indicates that the listener heard a transmission and, therefore, a word about the privacy of certain transmissions is in order. In general, most public service communications systems (not broadcast stations) such as police, fire, telephone, etc., are covered by Section 605 of the Communications Act of 1934 dealing with the secrecy of communications. The act only forbids divulging the contents of a transmission and has no effect upon listening to any station, even those who may be using some private transmission method. Therefore, reporting reception as such can cause no problems, but putting in the report some information

	S	I	N	P	O
		degrading effect of			
rating scale	signal strength	interference (QRM)	noise (QRN)	propagation disturbance	over-all readability (QRK)
5	Excellent	Nil	Nil	Nil	Excellent
4	Good	Slight	Slight	Slight	Good
3	Fair	Moderate	Moderate	Moderate	Fair
2	Poor	Severe	Severe	Severe	Poor
1	Barely audible	Extreme	Extreme	Extreme	Unusable

Fig. 5-3. The international SINPO code is often used to report the quality of a received signal.

about the content of the transmission is an unclear point, legally, as the law has never been subject to any real court test as far as casual listeners are concerned. It should also be noted that some states have laws against the installation of mobile receivers specifically designed for receiving police and-or fire department communications without obtaining prior approval for such installations. Again, the legality of such laws has usually not been questioned.

Receiving Antennas

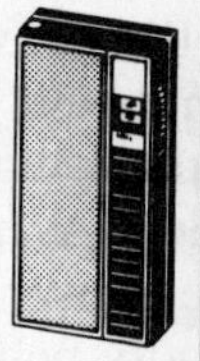

Chapter 6

Next to the quality of receiving equipment and operator skill, good reception is wholly dependent on an efficient antenna system. The requirements for a good receiving antenna system have a great deal in common with transmitting antenna systems, but there are a number of notable differences. The differences are usually not appreciated by most operators who believe that if they put up the best transmitting antenna possible and it correctly works with a transmitter, then it must also be equally good for receiving. The total response of an antenna used for receiving purposes as compared to the same antenna used for transmitting may be different because interference sources which do not affect the transmitting ability of an antenna drastically alter its response as a receiving antenna.

With the development of modern, high-sensitivity receiving equipment, the requirement for highly elaborate receiving antennas has been reduced. Although someone who has the space available may benefit from a large receiving antenna installation, amazing reception can be obtained with small antenna systems. The antennas described in this chapter are generally types that can be used over wide frequency ranges in the HF and VHF spectrum.

CHARACTERISTICS OF RECEIVING ANTENNAS

There are several important terms used to describe antenna characteristics which the reader should clearly understand.

Length. The length of an antenna is usually expressed in wavelengths. The length of antenna which can accommodate a complete wave must be at least one-half wavelength (λ) long since the wave travels in one direction along the antenna for one-half wave and then returns along its length for another half wave.

Polarization refers to the condition as mentioned in Chapter 1. If the wave which the antenna radiates is horizontal, the antenna is said to be horizontally polarized. Notice that it is the radiation polarization that is involved, not the physical orientation of the antenna. Most common antennas which are horizontal with respect to the earth's surface radiate horizontally polarized waves, but this is not always true.

Directivity refers to the directional characteristics of an antenna—the fact that it will respond more to (or radiate more) energy in some directions than in others. When interference problems exist while receiving stations from certain directions, an antenna with high directivity becomes very desirable.

Gain, a term which properly applies mainly to transmitting antennas, is a measure of the maximum intensity radiated by an antenna as compared to the maximum radiation from some reference antenna. The gain figure has no meaning unless the reference antenna is also stated. Dipole antennas or isotropic radiators (a hypothetical antenna which radiates uniformly in all directions) are usually used as references. For very small antennas (small in terms of wavelength), the directivity may be very high, but the antenna will have no gain.

Beamwidth refers to the width of the radiation pattern from the antenna at some given point. Since an antenna radiates in three dimensions, both horizontal and vertical beamwidths must be specified. The beamwidth may be taken at any point, although the usual points used are the half-power or 3-db-down points.

Impedance refers to the complex impedance at the terminals of the antenna at a particular frequency. Generally, it is desirable that the impedance remain purely resistive at the operating frequency so that it will directly match one of the common transmission line impedances of 50, 72, 300 ohms, etc. When this is not the case, some form of matching network can be used to bring about this condition.

Bandwidth refers to the frequency range over which an antenna can be used before some undesired conditions occur. The undesired condition may be an impedance in excess of some value which prevents efficient coupling between the antenna and the transmission line or it may be some undesired change in the directive pattern of the antenna.

SIMPLE DIRECTIONAL ANTENNAS

Before we discuss specific antenna types, it might be best to define what is meant by a "directional" antenna. Simply it is an antenna which concentrates its radiation in a specific direction as compared to a completely non-directional antenna which radiates (like a point source) energy in all directions. The idea of viewing an antenna's radiation in three dimensions is extremely important to understand this concept.

The dipole antenna, and what might be regarded as its variations, provides the basis for evaluating most simple directional antennas. The 3-dimensional radiation from a dipole can be depicted as shown in Fig. 6-1A. If this dipole were parallel to the earth's surface and you looked straight down at the antenna from above, it would have a so-called horizontal pattern as shown in Fig. 6-1B. If you were to look only sideways at the antenna, it would have a so-called vertical pattern shown as a circle around the wire. These horizontal and vertical patterns exist for a dipole an infinite height above the ground. In a real situation where the dipole is located a definite height above ground, the horizontal pattern will remain essentially the same but the vertical pattern will change markedly with the height above ground. For instance, Fig. 6-1C shows the vertical pattern for a dipole a half wavelength above ground. The "basic" dipole is usually taken as an antenna one-half wavelength long, although the name dipole simply indicates a 2-part antenna.

As was mentioned previously, a wavelength is equal to (in meters) 300 divided by the frequency in MHz. However, since a radio wave travels slightly slower in a conductor than in free space, and to account for the effect of insulators, etc., in the vicinity of the antenna, a slight correction of 2 to 5 percent is made. For wire or tubing antenna elements (up to 1 inch diameter), the following relationship is used for HF antennas:

$$\tfrac{1}{2}\lambda \text{ (in feet)} = \frac{468}{\text{freq. in MHz}}$$

The dipole is also a single-frequency antenna in the sense that the horizontal pattern shown in Fig. 6-1 exists at only one frequency and the center impedance value of approximately 70 ohms resistive is present only at the design frequency. At

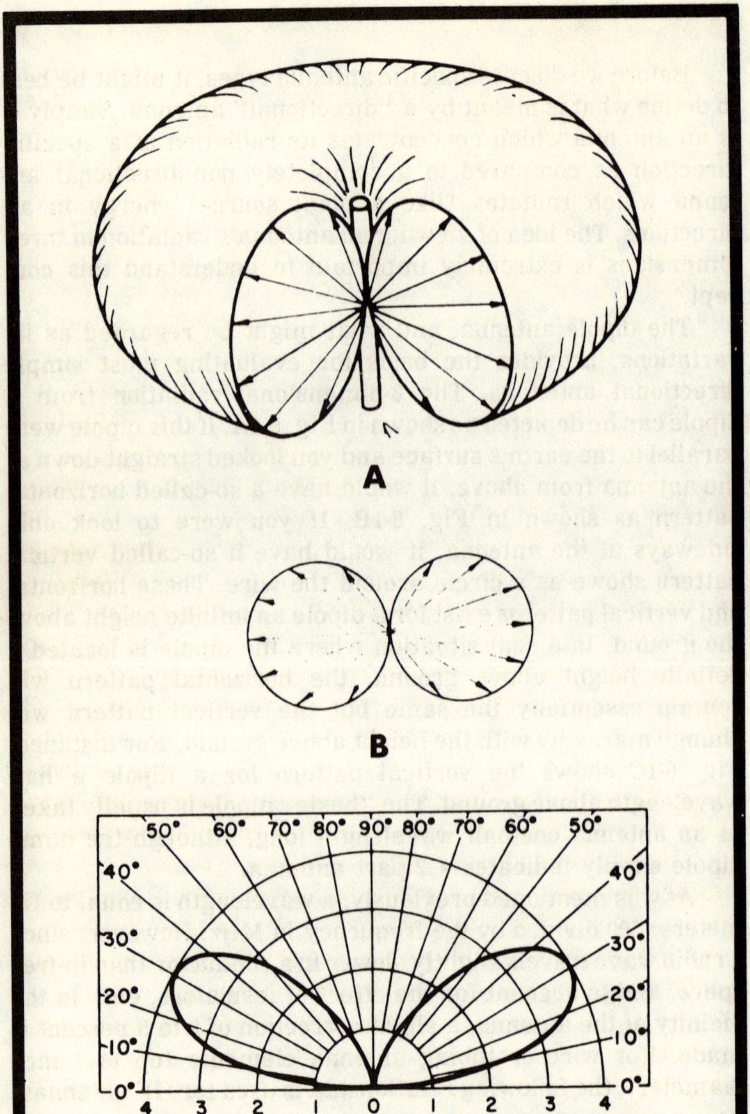

Fig. 6-1. In free space a dipole antenna's response pattern is three dimensional (A). The vertical and horizontal patterns (B) are the same in free space. In any practical situation where the antenna is elevated a specific distance above ground, the vertical pattern (C) takes on a specific response.

lower and higher frequencies, both the horizontal pattern and impedance will change. However, the impedance changes can be accommodated by various matching devices. A more important consideration, usually, for the receiving operator is whether the patterns that result on different frequencies are useful. On frequencies below the point where the dipole is a half wavelength long, the pattern will remain about the same, although the signal pickup will decrease. On frequencies higher than the half-wave frequency, the pattern will go through various changes, with the response being greater in specific directions than with a half-wave dipole. The main variation in response, if you refer to the horizontal pattern in Fig. 6-1B, is the gradual shift in response from a right angle position to the line of the dipole to the line of the dipole.

A dipole antenna one or more wavelengths long is frequently referred to as a long-wire antenna. They are definitely useful for many applications and, if long enough, provide some minor gain. The main disadvantage is that they possess no change in vertical directivity. That is, you could construct an extremely long long-wire of many wavelengths, but if it were elevated one-half wavelength, its main vertical response would remain about 30 degrees—certainly not the optimum for long-distance reception.

Simple half-wave dipoles can be parallel connected to a single transmission line, if desired, to preserve the basic dipole characteristic at each frequency for which a dipole is cut. At in-between frequencies, the response pattern will be a complex combination but still basically at right angles to the antenna. Fig. 6-2 shows the dimensions of a military type receiving parallel dipole, consisting of two dipole elements, designed to cover a large portion of the short-wave range.

The loop antenna is one of the most commonly used receiving antennas where limited space is available, particularly below 10 MHz. The basic loop (Fig. 6-3) may consist of one turn or multiple turns of wire either in the form of a circle or square. As long as the total length of wire in the loop is not greater than one-tenth wavelength it will have the response pattern shown. When the wire length exceeds one-tenth wavelength, other responses will appear which will destroy the deep side nulls in the response pattern. Because of these pronounced nulls, loops are frequently used for direction finding where ground-wave signals are involved. Loops are

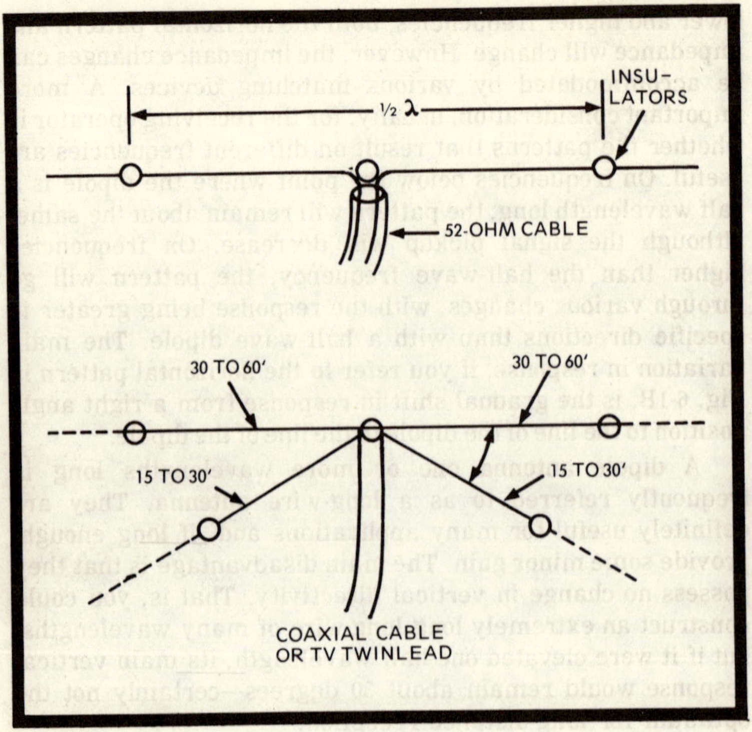

Fig. 6-2. Basic dipole (above), cut for one frequency, can be used for receiving or transmitting. A receiving-only dual dipole is shown below; it can be used over the entire short-wave band. The best response to signals is in directions broadside to the antenna (into and out of the page).

useful for rejecting interference where sky-wave transmissions are used, but not for exact direction finding because of the irregularities in the arrival of sky-wave signals.

Balance is very important in retaining the response pattern of the loop. Loops used in the presence of other metal objects are frequently covered with a metal electrostatic shield. The shield reduces effects from induction fields nearby, balances capacity effects and reduces precipitation static effects. The outer braid of a coaxial cable placed over a flexible plastic tubing (which encloses the loop wire turns) makes a simple but effective shield.

The loop should be tuned to resonate at the desired reception frequency for greatest signal pickup. This requirement reduces the usefulness of loop antennas to some degree unless some remote means is used to tune them over a range of frequencies. Motor-driven tuning capacitors have been used to do this. A loop about 12 inches in diameter with 4 to 20 turns of wire, as used on many aircraft, makes a convenient size for use on the LF to lower HF bands.

Ferrite rods can be used also for the construction of loop antennas. In general, such an antenna will have about as much pickup as a regular loop with the same diameter as the length of the ferrite rod. Except on special order, ferrite rods are available only in lengths up to 9 inches. However, several such rods may be bunched together to make a larger antenna.

Vertical antennas and similar variations are another type of simple directional antennas. If you imagine the dipole in Fig. 6-1A being turned vertically, it can be seen that directly off of the ends of the antenna (now vertical) there will be no response. However, there will be a form of all-around horizontal response. So, vertical antennas are frequently used for receiving purposes where an omnidirectional response in the horizontal pattern is desired. The response to low-angle signals is good, particularly when a good ground connection or system is used. The elevation must be kept low in regard to ground for best low-angle response or an artificial "ground

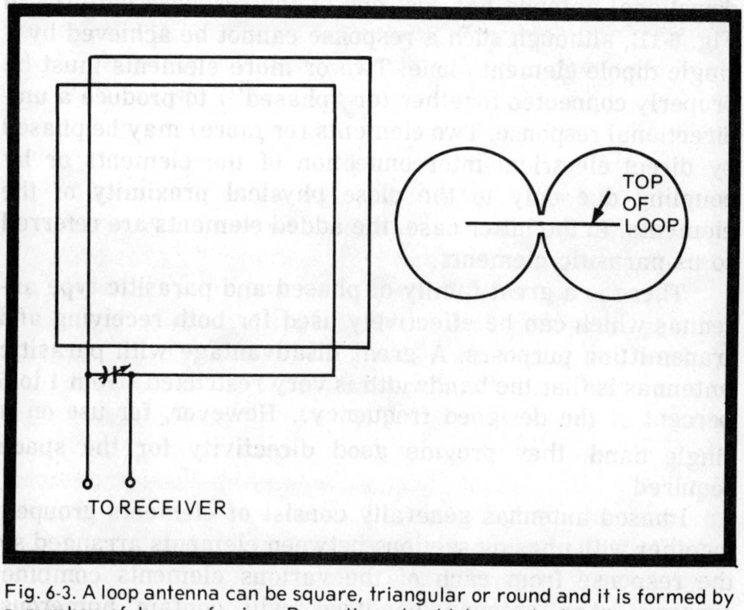

Fig. 6-3. A loop antenna can be square, triangular or round and it is formed by a number of turns of wire. Reception should be peaked with a variable capacitor.

plane" must be placed beneath the elevated antenna. The vertical half-wave dipole is frequently too long to be practicable. One solution is to use the ground or a group of 3 to 4 ground plane elements to effectively make up for one half of the antenna (ground plane antenna).

A short direct ground connection becomes very important with vertical antennas having no ground plane other than the earth. When these antennas are one-quarter wave long and used for transmitting purposes, a ground screen consisting of a mesh of wires a half wave or more long are frequently used as a so-called counterpoise. It may be buried or laid on the surface of the ground. In poor soil such a counterpoise should also be used for receiving purposes.

OTHER BI-DIRECTIONAL AND UNI-DIRECTIONAL ANTENNAS

The terms bi-directional and uni-directional mean an antenna which receives from two directions or only one direction, respectively. The simple dipole response in Fig. 6-1B is bi-directional. In more elaborate antenna forms, this bi-directional response becomes sharper in each direction. A uni-directional antenna has just one of the response patterns in Fig. 6-1B, although such a response cannot be achieved by a single dipole element alone. Two or more elements must be properly connected together (or "phased") to produce a uni-directional response. Two elements (or more) may be phased by direct electrical interconnection of the elements or by coupling due only to the close physical proximity of the elements. In the latter case, the added elements are referred to as parasitic elements.

There is a great family of phased and parasitic type antennas which can be effectively used for both receiving and transmitting purposes. A great disadvantage with parasitic antennas is that the bandwidth is very restricted (from 1 to 5 percent of the designed frequency). However, for use on a single band, they provide good directivity for the space required.

Phased antennas generally consist of elements grouped together with phasing sections between elements arranged so the response from each of the various elements combine properly. Any antenna handbook will contain numerous designs. Fig. 6-4 shows one simple design.

If half-wave dipoles are stacked, the vertical directivity is sharpened while the horizontal pattern remains essentially that of a simple horizontal half-wave dipole. Placing one half-wave dipole in the horizontal plane behind another half-wave dipole results in some sharpening of both the horizontal and vertical patterns. A multitude of arrays has been devised, based on combinations of the above forms.

Parasitic antenna combinations are distinguished by the lack of transmission line coupling between elements. When used for transmitting, the parasitic element is fed inductively by radiated energy from the center half-wave dipole or driven element, as it is called, since it connects to the transmission line. A basic parasitic beam is shown in Fig. 6-5. It may be used with both reflector and director elements or only one

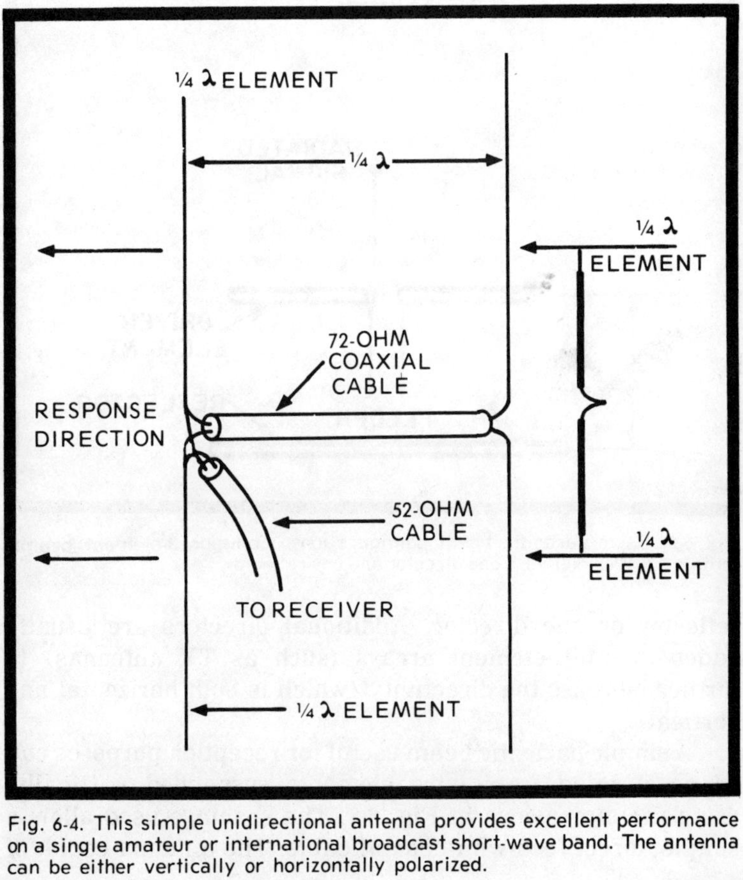

Fig. 6-4. This simple unidirectional antenna provides excellent performance on a single amateur or international broadcast short-wave band. The antenna can be either vertically or horizontally polarized.

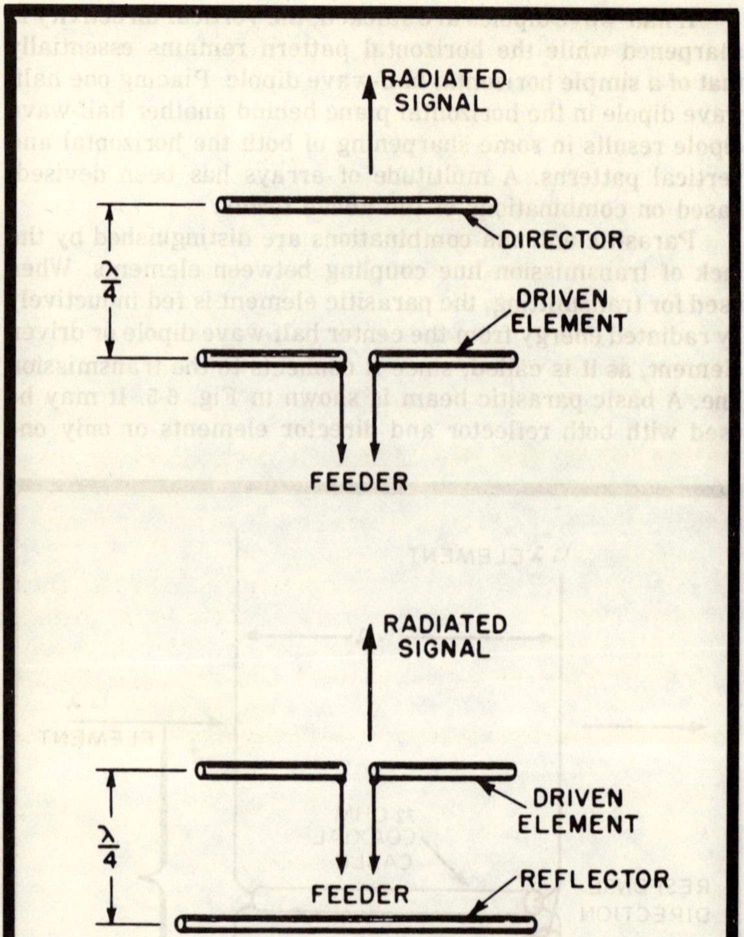

Fig. 6-5. Basic parasitic beam configurations. Common 3-element beams contain a driven element, one director and one reflector.

reflector or one director. Additional directors are usually added in multi-element arrays (such as TV antennas) to further increase the directivity (which is both horizontal and vertical).

A simple parasitic beam useful for reception purposes can be constructed from wire elements suspended vertically, using the dimensions in Fig. 6-5. The spacings used allow a simple, direct feed with coaxial cable. The antenna could, of course, be placed horizontally if supports are available. If not enough space is available for both elements, the director element should be eliminated.

ANTI-NOISE ANTENNAS

Atmospheric noise in the short-wave range can often be one of the most destructive forces against good reception. Such noise often originates in tropical regions through electrical storm activity and propagates like any radio wave in all directions. If you live north of the tropical area, one of the simplest measure to take is to construct or rotate an antenna so that it provides minimum response to the tropical area southward. This orientation may reduce the strength of the desired station (say from an easterly direction in Europe or Africa) somewhat, but if it reduces the atmospheric noise pickup even more, the overall result will be an improvement in reception.

You can also construct specific anti-noise antennas. These antennas are particularly useful as adjuncts to the regular receiving antennas when atmospheric noise levels become very high. A typical antenna of this type is shown in Fig. 6-6.

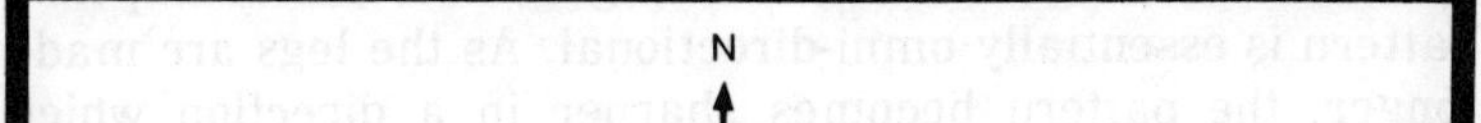

Fig. 6-6. Anti-noise or anti-static antenna. The antenna length is not so important as the fact that the long length of the antenna **must** be oriented in an east-west direction. Also, the height of the antenna **must** not be greater than 12 feet.

The orientation of the antenna and its approximate height must be as shown in the diagram if it is to function correctly. It cannot be erected on a high roof, for instance. The theory behind the antenna and its orientation is that it will respond weakly to low-angle noise signals arriving from the south (because of its low elevation) and also poorly but somewhat better to low angle DX signals arriving from an east-west direction.

EXTRA LARGE ANTENNAS

The hobbyist who is fortunate enough to have large amounts of real estate available can avail himself of the benefits of larger antenna structures as used by commercial installations. One of the simplest "large" antennas you can construct and one which can be expanded to almost any size is the "V" antenna.

The two dipoles or long wires in the form of a V as shown in Fig. 6-7 make a popular antenna because the response pattern is constant over a wide frequency range. It can be either bi-directional or uni-directional and, if the legs are long enough, provide reasonable gain.

When each leg is a half-wave long, the horizontal response pattern is essentially omni-directional. As the legs are made longer, the pattern becomes sharper in a direction which bisects the V angle. The antenna also possesses vertical directivity which is affected by the angle between the two legs. For best results, the antenna should be elevated one wavelength although it will still be effective if elevated as low as 0.4 wavelength. When the antenna is used over a wide frequency range, the usual practice is to make the angle between the legs suitable for the mid-frequency of the range.

The V antenna can be made uni-directional by means of terminating resistors of about 200 ohms connected from the far end of each V leg to ground. The resistors should be rated at one-third the transmitter power or, if used only for receiving, can be small non-inductive types of the 5- to 10-watt size. Although standard resistor values will give good results, significant improvement in the depth of the null to the rear can be obtained by adjusting the resistor values for minimum response to the rear. When unterminated, the input impedance varies widely with frequency. However, with a termination the input impedance over a wide frequency range will be the same value as the terminating resistance.

The rhombic is probably still one of the most commonly used antennas for commercial and military long-distance short-wave and lower VHF reception and transmission. It has broad frequency response, good directivity and is relatively simple to construct. Its main disadvantage is the space required. As with the V antenna, it can be made either bi-directional or uni-directional.

The usual form of a rhombic is the horizontal type shown in Fig. 6-8. It is capable of fairly high gain, particularly for the longer leg lengths. Since at least four variables (angle between legs, side length, height and type of termination) are

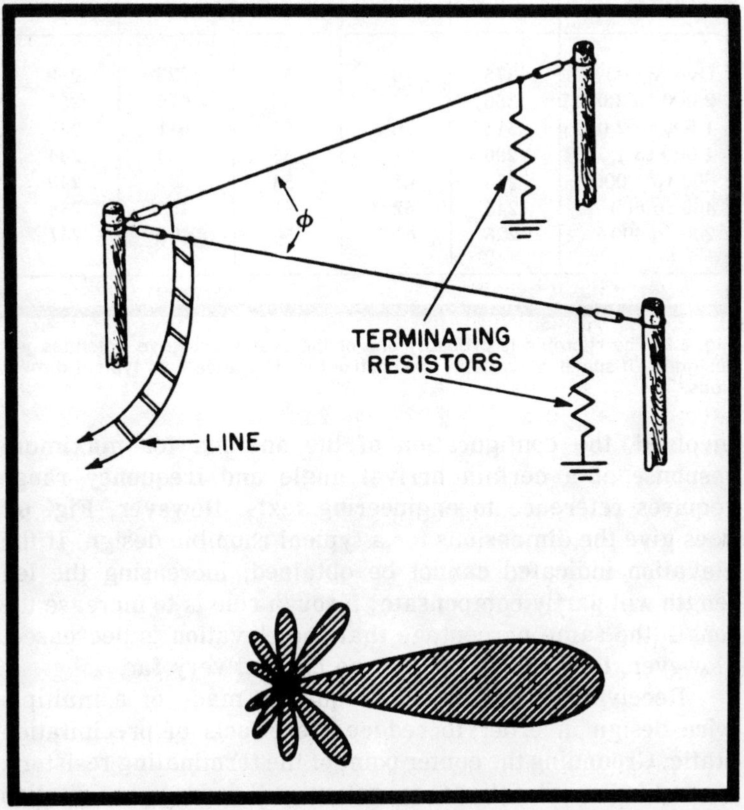

Fig. 6-7. A V antenna is uni-directional if terminating resistors (400 ohms each) are used, or bi-directional without the resistors. For short leg lengths (under two wavelengths), the angle between the legs should be 70 to 90 degrees.

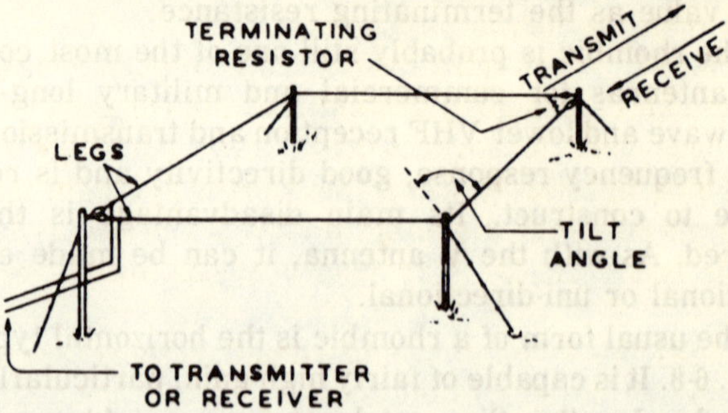

Range (mi)	Side length (ft)	Tilt angle* (deg)	Height (ft)	Length, end pole to end pole (ft)	Width, side pole to side pole (ft)
Over 3,000 . . .	375	70	65	723	268
2,000 to 3,000	350	70	60	676	251
1,500 to 2,000	315	70	57	611	228
1,000 to 1,500	290	67. 5	55	553	234
600 to 1,000 .	270	65	53	506	240
400 to 600 . .	245	62. 5	51	453	238
200 to 400. . .	225	60	50	407	237

Fig. 6-8. The rhombic is probably one of the best short-wave antennas yet designed—if space is available to construct it. The table lists typical dimensions.

involved, the configuration of the antenna for maximum response at a certain arrival angle and frequency range requires reference to engineering texts. However, Fig. 6-8 does give the dimensions for a typical rhombic design. If the elevation indicated cannot be obtained, increasing the leg length will partly compensate; a rough rule is to increase the length the same percentage that the elevation is decreased. However, this process cannot be carried very far.

Receiving rhombics are frequently made of a multiple wire design in order to reduce the effects of precipitation static. Grounding the center point of the terminating resistors, if used for uni-directional operation, will reduce precipitation static effects.

As with the V antenna, fine adjustment of the terminating resistor will produce significant changes in the depth of the null to the rear. When adjusted, the rear response can be 100 db less than the forward response.

Generally, when used over a wide frequency range, the rhombic can be designed around the mid-frequency. For instance, one used to cover 6 to 15 MHz would be designed around 9 or 10 MHz. However, no worthwhile gain for transmitting purposes for the effort involved will occur if the leg length is less than one wavelength at the lowest frequency used.

Besides the V antenna and rhombic, there are several large antenna designs which have frequently been employed for commercial and military receiving applications.

The Beverage or wave antenna is frequently used for long-distance reception on LF and MF and less frequently in the short-wave range up to 6 MHz. The antenna does not have to be highly elevated, but it must be at least two wavelengths long to be effective. It works best over poor soil. Basically, it is bi-directional along the length of the wire but is most often employed as a broadband uni-directional antenna by terminating one end. The backward null can be very pronounced when the terminating resistor is properly adjusted.

The "fishbone" antenna will give about the same reception performance as a large-size rhombic antenna. Its configuration, however, sometimes makes it easier to construct than a rhombic. It is terminated and used as a uni-directional antenna over a fairly wide frequency range.

The Wullenweber is one example of a number of antenna designs used in the short-wave range for direction-finding purposes. The output from each dipole of a multiple series is compared to the outputs from the other antennas to determine the direction of arrival of a radio wave. The main advantage of such systems is the very narrow beamwidths that can be developed for accurate direction finding as well as interference rejection. The Adcock system is similar, except that it requires no screen and its beamwidth is not as sharp.

Sterba and Bruce arrays where once used extensively, for instance, in the 1920s for transoceanic telephone service, but have now been generally replaced by other types such as rhombic and fishbone receiving antennas. However, if space is available they can be very effective. The main disadvantage probably is that they have a bi-directional response and there is no convenient way of making the response uni-directional.

The large antennas described here are classical ones and are presented in detail in any engineering text available in a library.

TRANSMISSION LINES AND COUPLERS

The purpose of a transmission line is to convey energy between an antenna and a receiver or transmitter. However, when a transmitter is involved, the antenna is the ultimate load for the transmitter and the objective is to transfer the output power to the antenna with minimum loss along the way. To do this the transmitter output circuit must be matched to the transmission line and the transmission line in turn matched to the antenna. When these conditions are not met, part of the power cannot be accepted by the antenna, non-uniform current distribution (standing waves) result on the transmission line and the rejected power radiates from the transmission line, heats the line or the transmitter output circuitry, or is otherwise uselessly dissipated. So, for transmitting, impedances must be correctly matched all along the way—a transmitter with a 50-ohm output working into a 50-ohm transmission line which is connected to an antenna with 50 ohms impedance at the transmitting frequency, for instance.

For receiving, the antenna is not the load but the source of energy and the system must be considered looking in the direction of antenna to receiver. The energy which the antenna absorbs from a passing radio wave must be passed on to the receiver with as little loss as possible. However, in addition, a minimum of unwanted energy must be allowed to enter into the system. The main source of the latter is usually noise picked up by the transmission line.

For maximum transfer of energy from antenna to transmission line, the impedances should match. However, at the receiver, the transmission line impedance may not match that of the receiver input circuit. Such a condition may be deliberate since it is not just the amount of signal but the signal-to-noise ratio which determines whether or not a signal can be detected. This ratio is affected by the coupling circuit in the receiver between the antenna and first receiver stage.

Basically, there are two types of transmission lines, balanced and unbalanced, corresponding to those antennas which are balanced to ground (a horizontal dipole, for instance) and those with one side connected to ground (a quarter-wave vertical, for instance). For transmitting purposes some leeway is often taken in using unbalanced transmission lines with balanced antennas (a coaxial cable with a dipole, for instance). However, for receiving purposes, such usage is not desirable since the noise energy is not balanced out.

Whenever possible, a transmission line which matches the antenna impedance should be used. A wide variety of balanced lines from 73 ohms to 450 ohms are readily available, and coaxial cable impedances range from 50 to 95 ohms. Generally, the attenuation or loss in most commercial cables is quite low throughout most of the short-wave range. However, at the upper end of the range and particularly where long transmission line runs are involved, care must be given to attenuation characteristics and a tradeoff made between low attenuation and cable cost.

Baluns effect a transition from a balanced to unbalanced (or vice versa) transmission line along with a 4:1 or no impedance transformation. Designs are commercially available which operate on broad-band transformer principles. They are effective over the entire short-wave range and provide balanced to unbalanced transition with or without impedance stepups.

Since the impedance of most antennas, except terminated types or other broad-band designs, vary considerably with frequency, so-called resonant transmission lines are frequently used to couple to such antennas. The resonant line acts much like the quarter-wave transformer, except that as the impedance varies at the antenna the impedance at the other end must also be varied correspondingly. This action is provided by some forms of antenna couplers. In practice, the transmission line length does not remain a quarter wave, but the complex impedances developed are matched by the antenna coupler's characteristics.

An antenna coupler can also serve the function of impedance transformation; for instance, to couple a 450-ohm balanced line (which is non-resonant) to a 50-ohm unbalanced receiver input. Antenna couplers are often useful in a completely non-resonant system (a 300-ohm balanced line connected to a 300-ohm balanced receiver input, for instance), particularly in the upper short-wave or lower VHF range. The receiver input coupling circuit at these frequencies is usually a compromise between many factors, such as signal-to-noise ratio and selectivity, and an antenna coupler will often provide improved reception.

ANTENNA SITING AND CONSTRUCTION

Selection and siting as well as construction of an antenna are all closely related. The selection of an antenna must be made according to the frequency characteristics desired as well as to the restrictions of the space available. Construction not only entails a stable mechanical structure but also one which insures minimum signal loss or noise degradation in getting the signal to the receiver.

The selection of an antenna must first of all be based on the frequency range and desired directivity. For instance, if you are interested in just general long-distance reception, a vertical half-wave dipole or ground plane antenna might be chosen. Usually, after a while, most operators become interested in reception from specific areas. Then, an antenna with some directivity is usually desired. The antenna actually used depends on a study of the characteristics of various antennas versus the space available for construction. The direction from which the signal will arrive from a distant station must be determined from a great circle chart. For antennas of sharp directivity, such as rhombics and very long V antennas, a more accurate map must be used. Since these antennas may have beam widths of only 10 degrees or so, an error of a few degrees either in establishing the map direction or actually in erecting the antenna, which is of a non-rotatable type, can greatly reduce the antenna performance.

When possible a fixed directional antenna should also be oriented by taking into consideration the origin of interfering stations and noise. The latter can be established by using a small but directive antenna, such as a short dipole or loop, with a receiver covering the desired reception frequencies and establishing the direction of noise arrival by utilizing the null responses from the antenna. Such measurements should be made over a period of at least several days, if possible, to establish a mean direction. With antennas of broad directivity or with multiple nulls in the response pattern, such measurements will enable you to site the antenna so minimum response occurs in the main direction of noise arrival while the main response in the desired direction is only slightly affected.

When determining antenna direction for short distances, such as with local ground-wave or line-of-sight reception, there is no need to use special maps. The commonly available Geodetic Survey maps for the U.S. are very useful for local or state coverage.

Another factor that you must establish is whether to use an antenna that responds to vertical or horizontal polarized waves. Radio waves that are refracted from the ionosphere generally exhibit a circular polarization regardless of the wave polarization at the transmitting site. Therefore, the receiving antenna polarization is not critical. When a choice is available, horizontal polarization seems to be favored, but there is no clearly established proof that it is better. Polarization does, however, become critical when local line-of-sight communications are involved. The receiving antenna must then be polarized the same as the transmitted wave or considerable signal loss will result.

Another factor to be considered in regard to polarization is the nature of the surroundings by the antenna. In heavily wooded areas, for instance, horizontal polarization will generally prove better (both for receiving and transmitting) than vertical polarization. The erection of the antenna in line with obstacles in the desired direction of reception should, of course, be avoided.

The construction of an antenna should be considered in two parts; the antenna proper and the transmission line to the receiver. The antenna itself, when used for receiving purposes, need not be constructed from any particular size of wire as far as signal losses are concerned. The size of wire used should be determined by mechanical considerations—keeping the wire taut and providing enough strength for security against weather conditions such as ice loading. Number 14 Copperweld (copper-covered steel) wire or stranded silicon bronze wire near salt-water areas are particularly suitable. Where the two wires that are insulated from each other pass over each other, egg-type insulators are preferred over the simple straight insulator with a hole at each end. The egg insulators cost little more, but if the insulator breaks the wires will still remain joined. If egg insulators are not available and straight insulators are used, a type made of Cycloac is preferred over the cheap, poorly formed porcelain types that tend to break easily when subject to tension. Tension on the antenna wires is easily maintained for antennas of mode-rate length by inexpensive compression type springs sold in boat supply houses.

When wires must be joined to each other to form a complex antenna array, the proper electrical bond between the wires becomes very important. Poor bonds which are not always noted when an antenna is used for transmitting will produce noise interference during reception. Soldering is the simplest means of bonding for the hobbyist but should never be depended on for mechanical stress. Small electrical clamps are useful for mechanical bonding.

The transmission line connection is a frequent weak point in many antenna installations. Open-wire line can be joined to a wire antenna by means of electrical clamps and soldering. Coaxial and other thin-wire lines must be more carefully joined because the conductors themselves can take only moderate stress and moisture must be prevented from getting inside the line. Home-built connectors which grasp the outer jacket of such cables can be constructed or purchased. To prevent moisture from entering, the end of the cable can be secured with an electrical tape and-or a coating of insulating lacquer.

The transmission line path from the antenna to the receiver should be carefully considered, especially in areas where many interference sources, such as motors or other electrical machinery, are present. Making the transmission line 50 or 100 feet longer to avoid the area of such interference will generally add little line loss as compared to the improved reception possible because of noise reduction.

When it is not possible to avoid interference areas, the use of a balanced and shielded transmission line should be considered. Another solution, less frequently possible, is to bury the cable a few inches in the ground.

A few precautions should be observed when using shielded or coaxial cables. Fresh cable of reputable manufacture should always be used. Bargain cable may be old and (particularly if long-stored military cable) contain a dielectric material which drastically changes its properties with age.

Moisture, bending and poor connections are the main installation faults to avoid. Moisture entering the cable will corrode the shield and destroy the cable's effectiveness. Bending with a radius of less than 10 or 15 times the cable diameter will tend to stress the dielectric to the point where it will crack as it ages. Poor connections will nullify the shielding properties of the cable as well as cause signal loss. Any large manufacturer will supply detailed instructions for the use of shielded cable and associated connectors; all instructions should be carefully followed.

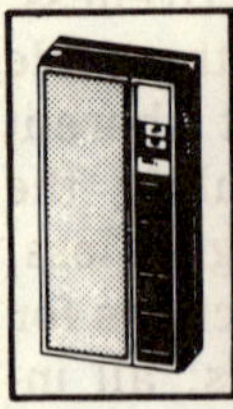

Chapter 7

Signal Modulation Methods

When tuning through the shortwave or VHF ranges, you are likely to encounter a confusing variety of signal types. Many will be easily recognizable such as the usual amplitude modulation used on the shortwave broadcast bands, FM as used on the VHF broadcast band, or single sideband (SSB) as used on the amateur radio bands or on some CB channels. There are many different forms of modulation for voice or data transmission which are more specialized and used only by commercial or military stations for reasons of economy, reliability, or privacy. You can hear these signals on ordinary receivers, but to detect them (convert the signal into understandable form) requires special accessories to a receiver. Nonetheless, it is fun to tune the various bands and encounter these signals even if you cannot detect the intelligence they convey. By developing some basic understanding of how modulated signals are generated, such tuning will be much more meaningful.

BASIC MODULATION SYSTEMS

All basic modulation systems relate back to the **carrier** of a radio transmitter. The carrier is simply a current generated by the transmitter and radiated as an electromagnetic wave from an antenna. The wave oscillates at a certain radio frequency. If one turns a CB transmitter on, for instance, without any microphone connected, what is radiated is the carrier. This carrier (Fig. 7-1A) is a sine wave and is referred to as a continuous wave or CW signal. CW is often used also to refer to what is more properly called a "keyed CW" (Morse code signal), but that is just a matter of laxity in the use of technical terms. Keyed CW is treated later as a modulation system. This may be thought by some as not a real modulation system; but, in fact, anything that is done to the basic carrier to make it convey intelligence is modulation.

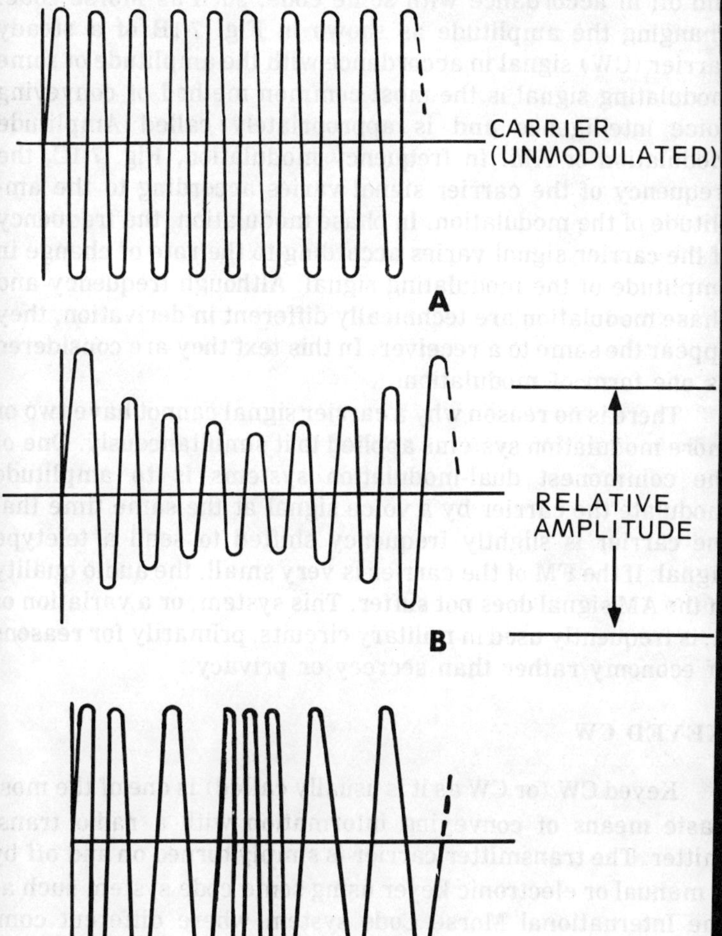

Fig. 7-1. Waveforms for various types of modulation.

A radio wave has four characteristics with respect to time and all modulation systems must vary one or more of these characteristics in order to convey intelligence. The four characteristics are presence, amplitude, frequency, and phase. Changing presence simply means to turn the signal on and off in accordance with some code, such as Morse code. Changing the amplitude as shown in Fig. 7-1B of a steady carrier (CW) signal in accordance with the amplitude of some modulating signal is the most common method of conveying voice intelligence and is appropriately called Amplitude Modulation or AM. In frequency modulation, Fig. 7-1C, the frequency of the carrier signal varies according to the amplitude of the modulation. In phase modulation, the frequency of the carrier signal varies according to the rate of change in amplitude of the modulating signal. Although frequency and phase modulation are technically different in derivation, they appear the same to a receiver. In this text they are considered as one form of modulation.

There is no reason why a carrier signal cannot have two or more modulation systems applied to it simultaneously. One of the commonest dual-modulation systems is to amplitude modulate the carrier by a voice signal at the same time that the carrier is slightly frequency shifted to send a teletype signal. If the FM of the carrier is very small, the audio quality of the AM signal does not suffer. This system, or a variation of it, is frequently used in military circuits, primarily for reasons of economy rather than secrecy or privacy.

KEYED CW

Keyed CW (or CW as it is usually called) is one of the most basic means of conveying information with a radio transmitter. The transmitter carrier is simply turned on and off by a manual or electronic keyer using some code system such as the International Morse Code system where different combinations of on-off sequences represent certain letters or numbers.

Keyed CW still remains one of the most important means of communication on the HF bands because of the simplicity of the transmitting equipment and because of its transmission advantage as compared to other systems. Manually keyed transmissions generally use the International Morse Code although there are a number of other codes used locally in various countries. There are also a number of international **Q** signals which allow commercial or amateur operators to communicate where language barriers exist or to abbreviate transmissions. A common usage is for ships entering foreign ports.

In general, manually keyed transmissions do not exceed 25 to 30 words per minute. Transmissions on the novice portions of the amateur bands may start as low as five words per minute. Machine sent transmissions extend up to 300 words per minute. Transmissions at the latter speeds may be recorded upon reception and played back at slower speeds for operator decoding or they may be decoded directly by electronic decoders. For privacy reasons, **burst** transmissions are sometimes used where a message is compressed in order to avoid having a transmitter radiating too long. A 20-word message, which a manual operator would take a minute to send at 20 words per minute, takes only four seconds when burst-sent at 300 words per minute.

AMPLITUDE MODULATION

The diagram in Fig. 7-1B shows how the carrier amplitude varies instantaneously when a modulating signal is applied. But the average carrier amplitude does not change, since it is increased and decreased equally. Also, its frequency does not vary. The energy that the modulating signal supplies appears in what are called side-bands. That is, when two frequencies are combined in a non-linear circuit (one where resistance varies according to the amplitude of the applied voltage) such as a vacuum tube, the output consists of the original frequencies plus sum and difference frequencies plus multiples of the original and sum and difference frequencies. However, for a properly designed transmitter only the original and sum and difference frequencies are important. Also, since the lower modulating frequency cannot pass through the transmitter output circuitry, only the carrier plus sum and difference frequencies are significant.

While Fig. 7-1 is a plot of amplitude versus time, Fig. 7-2 is a plot of amplitude versus frequency. Note that two side-band signals appear for each modulating frequency. They are removed from the carrier frequency by a frequency equal to the

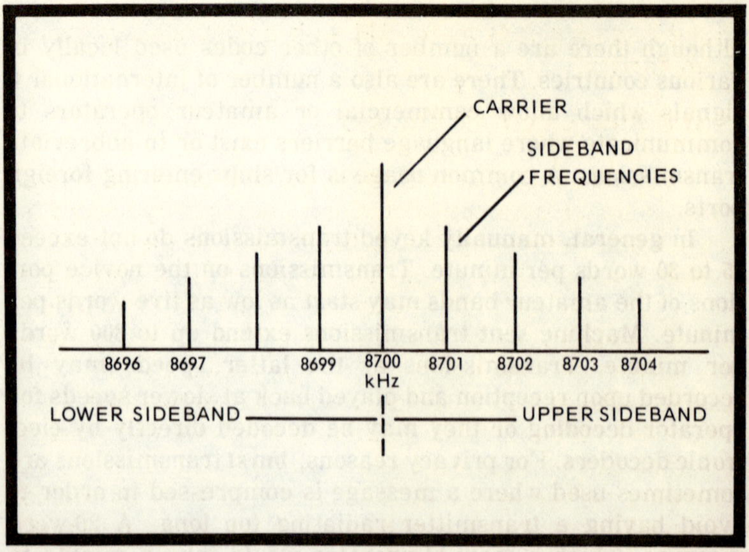

Fig. 7-2. An AM signal with its two sidebands can easily occupy a total frequency space of 6 to 8 kHz.

modulating frequency and are of an amplitude directly proportional to the amplitude of the modulating signal. The bandwidth of the total modulated signal is twice the frequency of the highest audio or modulating frequency used. The bandwidth of most domestic and international broadcasting stations is 10 kHz, while stations only interested in speech communications will have bandwidths of 5 to 6 kHz. AM can also be called a double side-band signal with full carrier.

Imagine a receiver tuned across an AM signal. If the receiver bandwidth could be made small enough, individual signals could be picked out. Some receivers do just this in the case of the carrier frequency. Since this signal is always present, while sideband frequencies may or may not be present according to the modulating signal, it can be used for control purposes. The carrier frequency can be used, for instance, to automatically make the receiver frequency track with the transmitter frequency.

The main disadvantage of AM, as far as a transmitter is concerned, is that it must constantly supply a carrier signal whether it is being modulated or not. Some transmitters are of the controlled-carrier type where the carrier is only turned on when a modulating signal is present. Such a signal will sound like ordinary AM on a receiver, although the receiver S meter will not remain steady as for AM with a continuously transmitted carrier.

SINGLE SIDEBAND AM OR SSB

In Fig. 7-2 notice that both the upper and lower sidebands of the AM signal are identical and only one is really needed to convey whatever intelligence is being sent. Single sideband transmissions are half of an AM signal. Filters and other circuitry remove one sideband and all or part of the carrier.

A SSB signal can be transmitted with full carrier, partially suppressed carrier or a fully suppressed carrier. The signal with full carrier is sometimes called "compatible AM" because any receiver which receives normal AM will receive this signal. Since the carrier signal is still present, there exists a reference for comparing the sideband signals to determine which modulating frequency they represent (demodulation). However, the transmitter must still supply most of its power to the carrier signal.

If the carrier amplitude is reduced but not eliminated (20 to 30 db from its full value), effectively more power can be put into the sidebands. However, the carrier is no longer strong enough to be used directly as a reference for demodulation. To replace it, a local carrier must be generated in the receiver. This local carrier can be generated using the suppressed carrier as a reference or can be self-generated such as with the beat frequency oscillator in a receiver. The suppressed carrier can also be used as a reference to control the receiver tuning so that it locks onto the transmitter frequency. Overseas telephone service, broadcast relay stations and an increasing number of maritime and aeronautical stations use this method of transmission.

If the carrier is completely eliminated (reduced 50 to 70 db for practical purposes), all of the transmitter power can be put into the sideband. A local carrier must be generated for demodulation, although there is no simple way to reference its frequency to the transmitted signal. The signal is tuned in by aural means. This method of transmission is most commonly employed by amateur operators, although it is gaining usage for commercial and military communications also. The latter is due to the extremely stable tuning and frequency control methods which have been developed in recent years which make the need for a pilot carrier unnecessary.

For the receiving operator, the main advantages of SSB are that a receiver with only half the bandwidth is necessary to receive the same voice signal as with normal AM and fading effects are reduced. The latter is due to the fact that under very selective fading conditions, the carrier of an AM signal may fade more than the sideband signals. Since a SSB receiver generates its own carrier for demodulation, these effects are not present.

FREQUENCY MODULATION

In frequency modulation, the amplitude of the modulating frequency determines the change or deviation of the carrier frequency from its unmodulated value. Unlike AM, this deviation can be made almost any desired amount for a given modulating amplitude. For instance, a 1,000 cycle tone of 1 volt amplitude is set to shift a carrier 15 kHz from its resting position. A 4,000 cycle tone of 1 volt amplitude will also shift the carrier 15 kHz. However, in the first case the carrier makes this shift 1,000 times a second, and 4,000 times a second for the 4,000 cycle tone.

One often encounters the term narrow-band and wide-band FM. There is no exact definition of these terms, but narrow-band refers usually to a small frequency shift with modulation (5 to 15 kHz) while wide-band refers to greater shifts. FM may also set up multiple sidebands for one modulating frequency. However, unlike AM, the sideband products beyond the primary ones are not insignificant. The amplitude of these sidebands is a function of the frequency deviation and the modulating frequency. Another significant feature of FM is that the carrier amplitude varies with modulation since the energy to form the sidebands comes from it.

The advantage of an FM signal with a large deviation (75 kHz for FM broadcasting) is the immunity to noise which the receiver circuitry can provide (since noise is an amplitude variation to which the receiver will not respond). In fact, noise of an amplitude only slightly weaker than the signal will not be heard.

Because of the limited spectrum space available in the 30 to 50 MHz range, police, military, and industrial stations operating in this range generally use a 15-kHz deviation.

Below 30 MHz so-called narrow-band FM is often used. There is no transmission advantage to the use of narrow-band FM over AM, but it does simplify transmitter circuitry.

CW TONE KEYING

Tone keying is the same as CW keying except that instead of the carrier being keyed, a tone is keyed which feeds either an AM or FM transmitter. The only advantage to this method is that the receiver used need not contain a beat frequency oscillator. However, the signal occupies considerably more bandwidth than a CW keyed signal. When used with an AM transmitter, such transmission is often called modulated CW (MCW). In its simple form (keying one audio tone), such transmissions are still frequently used in maritime broadcasts, navigational aids, and some military stations.

FREQUENCY SHIFT KEYING OR RADIOTELETYPE

In frequency shift keying the frequency of a transmitter's carrier is shifted up and down from its control or resting position, or a single audio tone's frequency is shifted as transmitted by an AM or FM modulated transmitter. The shifting of frequency might be discrete or continuous.

For example, in facsimile picture or weather map transmission, the shifting is continuous. A drum containing the printed matter, usually a weather map or news photograph, is scanned by a photoelectric cell which registers the darkness or lightness of a very small area. This variation is used to shift the frequency of an audio tone between, for example, 1200 Hz for black to 3600 Hz for white. This continuously shifting audio tone can be fed into the output of any AM or FM transmitter designed for voice operation. Another variation is to use the photoelectric cell output to directly shift the carrier frequency of a CW transmitter with a black to white frequency swing of about 1500 Hz.

A variation of the above scheme used to transmit text material on a narrow tape was the once secret German Hellschreiber. However, the signal was not generated by scanning but rather by a keyboard type device which always sets up the same signal sequence for a given letter. Like facsimile, which might require 15 minutes to transmit one

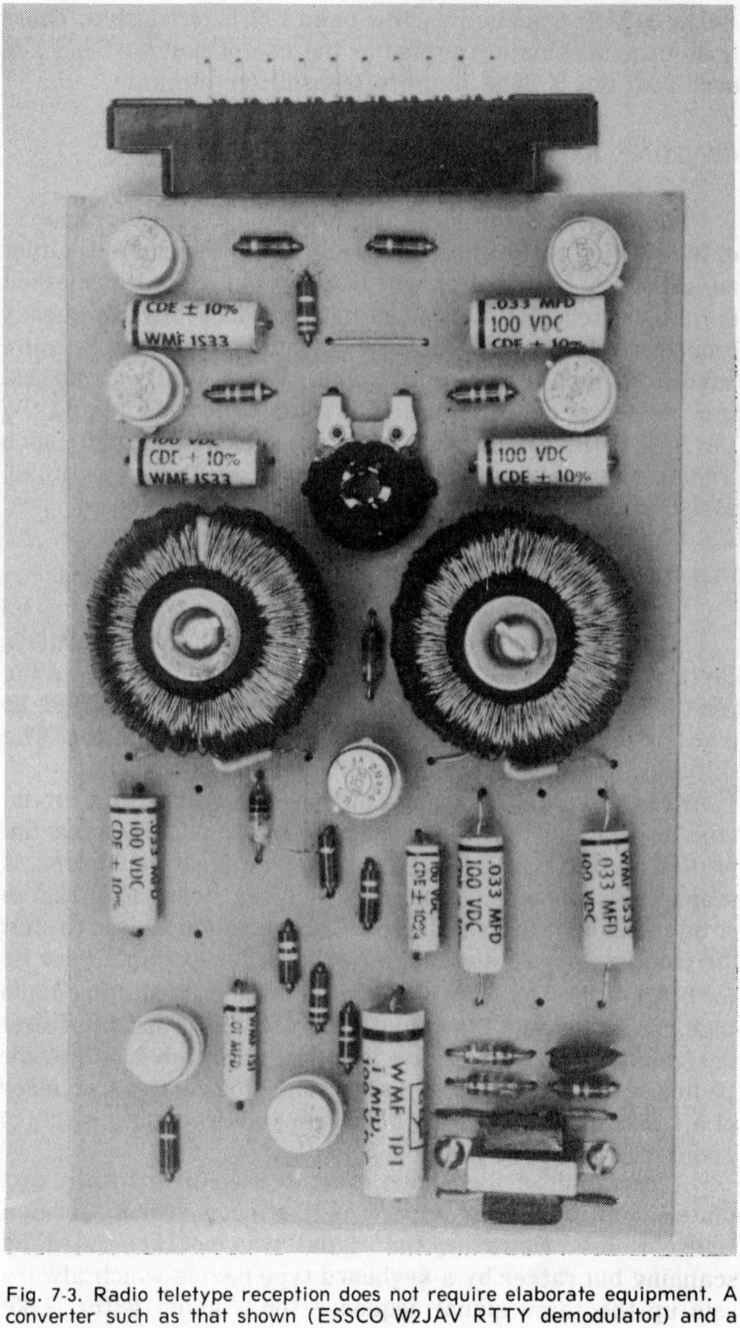

Fig. 7-3. Radio teletype reception does not require elaborate equipment. A converter such as that shown (ESSCO W2JAV RTTY demodulator) and a surplus teleprinter machine are the basic accessories needed.

page of material, it is a very slow method of transmission. The transmission of text material on the HF bands is now done almost entirely by teletype.

Teletype transmission is an example of discrete frequency shift transmission. Almost all single-channel teleprinter transmissions on the shortwave bands are accomplished by shifting the carrier frequency of a CW transmitter between two frequencies separated by 850 Hz. The teletype code is similar to the Morse code but designed for machine operation. It consists of a start pulse, five coded pulses for each letter or symbol, and a stop pulse. The time duration of these pulses will vary depending upon the speed of transmission. Amateurs use the 50-speed code while commercial stations may use 60, 75 or 100 speed. Other shifts beside 850 Hz are used; 170 Hz being one commercial standard, especially on the LF bands.

Audio frequency shift keying is also sometimes employed for teleprinter operation, especially on VHF where accurately shifting the carrier frequency may require too elaborate circuitry. In this case, the two audio frequencies 2125 and 2975 Hz are usually employed for the mark (on) and space (off) pulses.

A frequency-shifted signal actually allows for two methods of reception. The receiver may be set up to respond to both the mark and space frequencies; the change between the two frequencies indicating the information being sent. In this manner, noise interference is suppressed. On the other hand, if interference from another station exists on the frequency used for mark or space, the receiver bandwidth can be reduced to receive only one of the two frequencies since each frequency contains the same intelligence. However, noise will then more severely affect reception.

The hobbyist who owns a good shortwave receiver may be interested in adding radioteletype reception. Actually, it is one of the few specialized modes of reception that may be added for relatively moderate cost and opens up an interesting world of reception of news services, weather reports, etc. not available to the general public. A teleprinter machine is, of course, required. A number of such printers may be obtained quite reasonably on the surplus market. The only other equipment required is a frequency-shift converter such as that shown in Fig. 7-3. Such units are reasonably priced ($25 to $50) and convert the received frequency shift signals into DC pulses to operate the teleprinter. A specialized text on radioteletype reception should be consulted before purchasing the actual equipment; but, with care, the added capability of radioteletype reception to a sensitive, stable shortwave receiver is not beyond the technical capability of the average hobbyist.

MULTIPLE-CHANNEL SYSTEMS FOR VOICE AND DATA

For reasons of economy, many systems have been devised to transmit several independent intelligence channels over one transmitter's carrier. The receiving operator, unless he has some specialized receiving equipment, will normally not be able to receive these transmissions. However, he will certainly hear them in many portions of the shortwave and VHF bands and so a brief review of their characteristics is presented.

Multiple-Channel Keyed CW and Teleprinter

The simplest multiple-channel keyed CW transmission possibility is to key a separate audio tone for each channel desired and feed this signal into a conventional AM or SSB transmitter. The on-off keying of individual tones fed into an SSB transmitter by teleprinter signals is one of the most common methods of multi-channel radio teleprinter transmission. Up to 16 teleprinter channels may be transmitted in the 3 kHz bandwidth normally occupied by one voice channel. The tones are spaced 120, 150 or 170 Hz apart in different commercial systems. Reception of such a transmission is basically no more complicated than that for a regular teleprinter signal, provided the receiving equipment can adequately separate the various tone frequencies through proper selectivity.

Although on-off keying is normally used, each of the tones could also be frequency shifted over a very narrow range by the teleprinter signals. Either method, because it divides a given frequency bandwidth into a number of independent channels, is referred to as frequency-division-multiplex.

Another method of multiple-channel transmission is time-division-multiplex. In this method, time, instead of frequency, is divided. For instance, sending at 240 words per minute is equivalent to four 60 word per minute transmissions. So four 60 word per minute signals are scanned in turn at the transmitter at a 240 word per minute rate and transmitted at this rate. At the receiving end, a scanning device synchronized to the one at the transmitting site, redistributes the individual teleprinter signals. The transmitted signal would appear the same as any teletype signal except its speed would appear excessively fast and one could not detect the transmission without a special receiving adapter.

Multiple-Channel Voice Transmission

The most common method used on the HF bands for dual-channel voice transmission is independent sideband (ISB). Since SSB transmission leaves one sideband unused, another voice channel can be placed on this sideband. Refer to Fig. 7-2 and imagine a separate voice channel exists on either side of the carrier. Of course, this means that the total power available from the transmitter must be distributed between two sidebands rather than concentrated in one sideband. This method of transmission is often used by overseas telephone companies (with speech privacy equipment) and by broadcast relay stations. Generally, only receivers capable of normal SSB reception can receive such signals, although only one sideband can be received at a time; whereas commercial receivers would receive both channels simultaneously.

Another method of placing two to four voice channels on an ISB transmitter is by means of frequency division multiplexing as illustrated in Fig. 7-4. More than four channels are not used in the HF range because of the lack of frequency space.

Fig. 7-4 shows part of a four channel voice system: that is, two voice channels are shown on one sideband. The other sideband could also contain another two voice channels. But, as is often done, one sideband may also be used for teletype signals. In the illustration shown, the one sideband contains 16 teletype channels via tone keying. In this manner, a tremendous amount of traffic, both voice and data, can be handled by one shortwave transmitter.

Detecting or demodulating a complex signal such as four channel voice is complicated technically but not as complicated in hardware as one might imagine. To receive such a

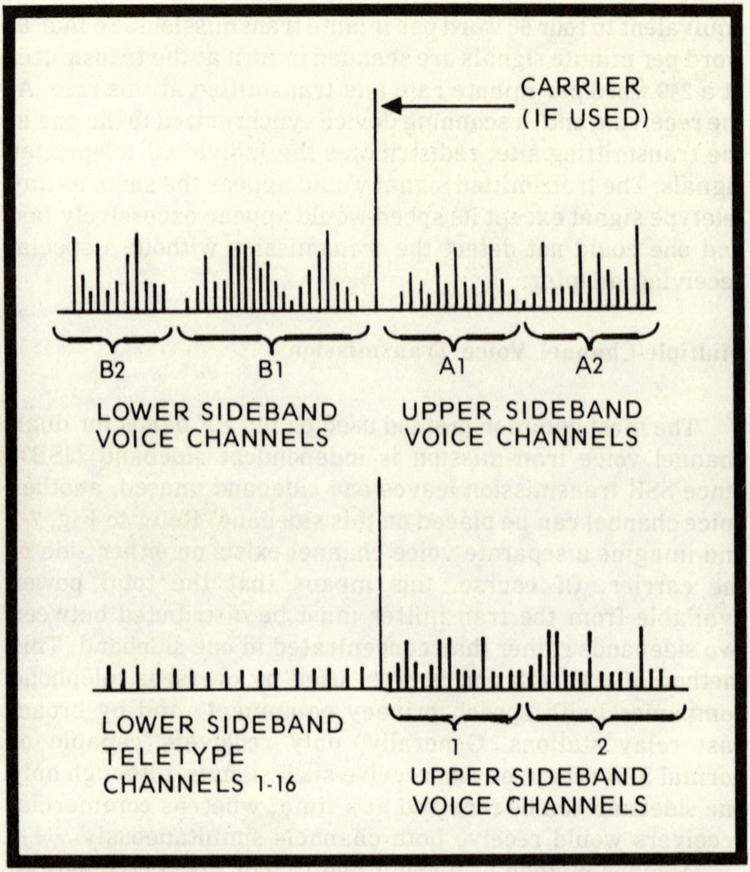

Fig. 7-4. Commercial stations often make elaborate use of one transmitter by filling each sideband with multiple voice or data transmissions simultaneously.

multiple-channel signal a good, stable receiver (preferably crystal-controlled) is necessary. In addition, a special demodulator unit such as that shown in Fig. 7-5 is necessary. Many such commercial units are available.

Narrow-Band Voice Transmission

Future years will see much increased use of narrow-band speech equipment with voice transmission in bandwidths as small as 100 or 200 Hz. The reduction of the normal voice bandwidth of 300 to 3,000 Hz is accomplished by equipment known as **vocoders**. They operate on the principle that at any

Fig. 7-5. Commercial equipment such as that shown for four voice channels is used together with a receiver to separate the complex signals of Fig. 7-4.

one time in speech production, energy is present only in certain portions of the 300-3000 Hz range. For each main energy band into which speech can be directed, two analog signals are generated, one relating to the center frequency of the channel and another to the amplitude in the channel. These outputs are digitalized and then each output scanned in turn to produce a long pulse train, which can be transmitted the same as any data signal.

The analog signals, each of which is a relatively low frequency of 20 to 40 Hz, can also be frequency translated next to each other and fed into the audio input of any AM or SSB transmitter, thus allowing voice transmission in only a few hundred cycles bandwidth. Vocoder equipment is now mainly used by the military services. Commercial and amateur use will develop eventually as equipment designs improve.

Privacy and Secrecy Modulation Systems

The terms **privacy** and **secrecy** in radio use are not too carefully defined. One can say, however, that privacy transmission refers to communication of such a nature that anyone may hear the signal being sent but cannot demodulate or detect it. Secrecy refers to a method of transmission whereby an attempt is made to keep even the presence of the transmission secret (such as burst CW transmissions as described before). Highly sensitive transmissions often employ a combination of privacy (coding) plus secrecy transmission systems.

Systems are used to provide privacy or secrecy in transmission for reasons ranging from commercial to military interests. Privacy systems differ, of course, from abbreviation codes used for commercial reasons where a group of letters might stand for a standard phrase, sentence, etc., (for example: Bentleys and Peterson business codes).

The coding of a message may be done either before it is transmitted or simultaneously with its transmission. Morse code messages are generally coded before transmission in any one of various systems whether the message is sent by electrical or physical means. Teleprinter messages can be encoded either before or during transmission.

The basis for most teleprinter coding systems is very simple. The plain message is compared, pulse for pulse, with a coding pulse. When both pulses are the same, a mark pulse is sent and when they are dissimilar a space pulse is sent (or a complicated combination thereof). The same coding pulse would be used to decode the message at the receiving end. No simple coding pulse would be used, of course, and the operation would be mostly electronic, the objective being to change the code pulse in a pseudo-random fashion so that study of the coded messages gives no clue as to any regularities in the plain message or code pulse. The same general scheme can be used, of course, to encode data transmissions, including speech transmissions converted into data form.

For applications which require speech privacy rather than secrecy, inverted speech or scrambling systems are frequently used. The simplest system is to invert the entire voice range, as shown in Fig. 7-6. Here, 300 Hz is transmitted as 2700 Hz and 2700 Hz as 300 Hz. Standard commercial inversion equipment divides the speech range into 3 to 5 parts (300 to 900 Hz, 900 to 1500 Hz, and 1500 to 3000 Hz, for example). These segments are placed, according to a peridoically changing code, into some segment of the 300 to 3,000 Hz transmitted bandwidth. The individual segments may also be inverted. Other variations involve providing for dividing the speech range into 2 (or more) parts, inverting them, and then making the dividing point vary continuously according to some fixed scheme.

Jamming Signals

Jamming transmitters use conventional forms of modulation, but, because so many are heard on the shortwave bands, a short summary of their characteristics is presented. Of course, what constitutes jamming is a moot question. If Radio Moscow sets a transmitter broadcasting a conventional program on the same frequency as Radio Tirana uses, is it jamming?

Conventional broadcasting material is not the most effective jamming signal, however. There are a multitude of specially developed jamming signals but the most frequently heard (against voice transmissions) are:

1. Pulse-keyed CW which sounds somewhat like an electric drill or other high-speed motor.

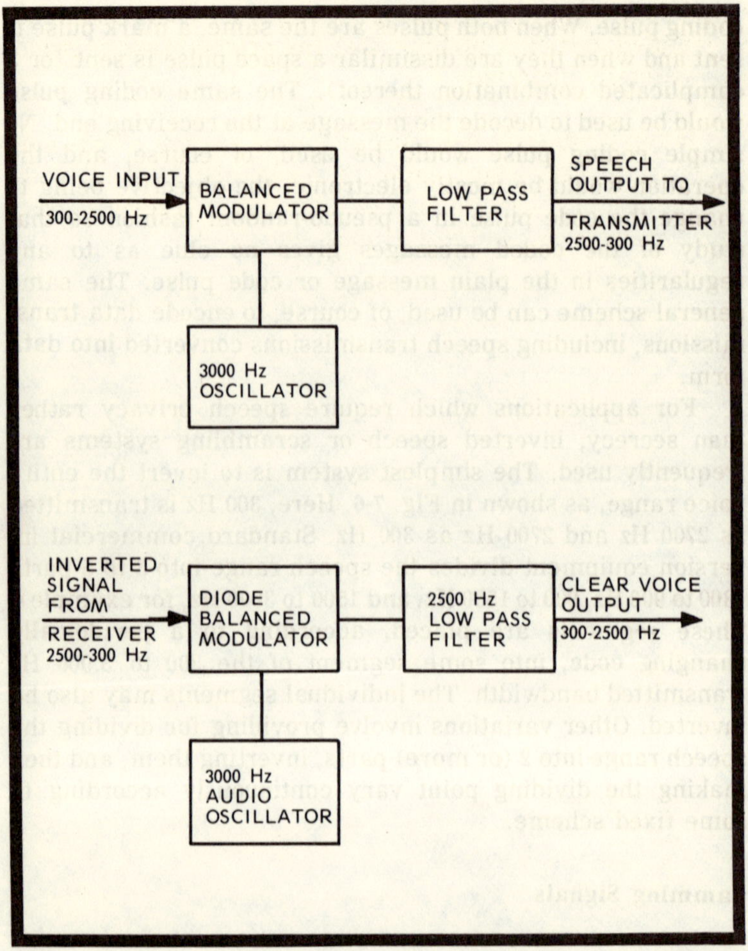

Fig. 7-6. Simple speech inversion for privacy transmission is often used in the basic form shown or with more elaborate variations.

2. Random noise AM which sounds like a tremendous increase in atmospheric noise.

3. Noise-burst, which is similar to the preceding type except that the noise occurs in random bursts.

4. Carrier-sweep which involves frequency shifting a carrier at a rapid rate. The sound in an AM receiver is somewhat like an automobile engine at high speed.

5. Grunting which is produced by modulating an AM transmitter with a very low audio frequency varying at a random rate. It sounds the same as it is named.

6. More advanced techniques involve reception and rebroadcasting, by the jamming station, of the transmission to be jammed in such a manner that the original transmission sounds totally distorted.

The latter method is not normally used to jam shortwave broadcast transmissions but it is a common method, with many variations, used to jam military voice transmissions, data transmissions and even radar signals.

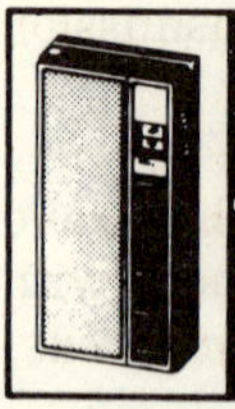

Chapter 8

Interference

Interference in reception becomes a problem only if the interference level is above that which can be tolerated to properly receive a signal. The level that a signal must be above the noise level in order for information to be derived from a signal depends upon the type of modulation and demodulation. For high-quality AM broadcast reception with a simple home receiver, the signal must be 30 to 40 db above the noise level. For international AM broadcasts, SSB, keyed CW, etc. the signal must be from 3 to 15 db above the noise level using modern HF receivers.

An interference problem, therefore, may or may not exist in the same location depending upon what type and quality of reception is required. **Interference** can also be classified as to whether it is internal or external. Many receivers themselves produce interference signals or, because of their construction, allow the easy entrance of external interference signals.

An enclosure for a receiver should provide not only environmental protection but also electrical isolation. The environmental protection needed naturally depends upon where and how the receiver is to be used. For military requirements, severe shock and vibration resistance may be required. Climatic conditions, particularly high humidity and temperature, may require construction with varnish fungus protection; salt water environments require the use of noncorrosive metals, etc. However, for normal use, adequate structural rigidity and ventilation is provided by formed sheet metal cabinets.

Electrical isolation to restrict the signal input to the antenna terminals is not always adequate with simple cabinets. The isolation of an enclosure can be checked by terminating the antenna terminals inside the enclosure in a resistance equal to the receiver's nominal input impedance (a simple 50-ohm resistor across the antenna terminals). The receiver is then placed in a known shielded environment (such as a shielded room or deep basement) to record any internally generated spurious signals and background noise level. The receiver is then tuned through its ranges in its normal environment. Any appreciable additional signals or noise heard are then due to signals entering the receiver through some path other than the normal antenna input circuit—direct pickup through perforations in the cabinet or via external wire connections (power line, audio output line, etc.). If the receiver is normally used in an area where a transmitter is operated simultaneously, the transmitter should be energized during this test.

If external problems are found by this simple test, perforations in the cabinet itself can be covered by a copper screen mesh bonded for good electrical connection to the cabinet. Other longitudinal openings can be shielded by using Belden **finger stock** across the opening, being careful that good electrical contact exists by removing any insulating paint, etc. where the finger stock makes contact. The bonding of the various parts of the cabinet to each other should also be carefully checked. Mechanical rigidity does not necessarily insure good electrical bonding between sections.

TYPES OF INTERFERENCE

The sound of noise in a receiver can sometimes be used to identify and possibly even to localize the source. A strong hum may indicate antenna lead pickup from a power line either inside or outside a structure. An improper ground may also produce hum even though the ground connection appears clean and tight. This can happen if the ground connection is too long or indirect or if the ground lead is common to some noise source. A buzzing sound is usually produced by a device which rapidly makes and breaks an electrical contact such as electric razors, vibrators, and certain power tools. When the speed with which the contact is made is slower, the sound will be more that of a clicking or shot type. Possible sources for such interference are automobile ignitions, teleprinters, and display signs. The same type of sound, but occurring irregularly, could indicate relays, elevators, power switches, traffic signals, or heating systems controls. A sizzling type of sound, whether of short or prolonged duration, may be caused by a neon light, diathermy machine, poor power line connections, or, if transmitting equipment is active in the area, loose contacts or poor bonding in antenna and grounding systems.

TRACKING INTERFERENCE SOURCES

If the sound of the interference gives no clue as to its identity, a battery-operated receiver can be used as a locator. Preferably, a receiver covering the same frequency range as the station receiver should be used, with its AVC circuit turned off and with an antenna consisting of a few inches of coaxial cable with the center conductor exposed one or two inches. Alternatively, a simple transistor AM broadcast band receiver can be used with its built-in antenna, if the interference appears to be spread across the entire frequency coverage of the main receiver. It cannot be used as an effective exploring device if the interference is confined to only a certain frequency range (above 20 MHz, for instance). Suspected noise sources are approached and any increase in noise interference noted until the offending source is found. If possible, the suspected source should be shut off as confirmation that it is the cause of interference. The tracing of noise interference to a supposed source may only in turn lead to another source. For instance, Fig. 8-1 shows how the actual interference source may be completely remote from the receiving location but through coupling appears to originate within the receiving location. Such a situation is frequent in apartment houses and industrial locations and the tracing of the true source of interference can become a tedious although not impossible task.

The situation is considerably simplified in a separated or isolated location such as a private house. Often, it is clearly obvious as to what the causes of interference are, as when heating systems or electrical appliances turn on. If the source is not obvious from correlation with the start of operation of various electrical devices, all electrical circuits should be switched off systematically until the source is located. Usually this is the only way to locate the odd case of interference caused by a loose electrical connection somewhere in the wiring system.

Another clue to locating the source can be gained from determining by what path the interference gets into the receiving system. If the receiver antenna terminals are terminated and interference persists, it may be entering through the AC power line (or the cabinet, if it is not well shielded). If no interference is noted with the antenna terminals joined through a resistor, the resistor should be removed and the antenna transmission line should be connected with its far end terminated in the line characteristic impedance by a small ½-watt resistor. The line should run in the same position as it does when connected to the antenna. Interference pickup will then indicate that the source is entering the system somewhere along the transmission line run. Some slight increase in noise pickup is to be expected since no line is per-

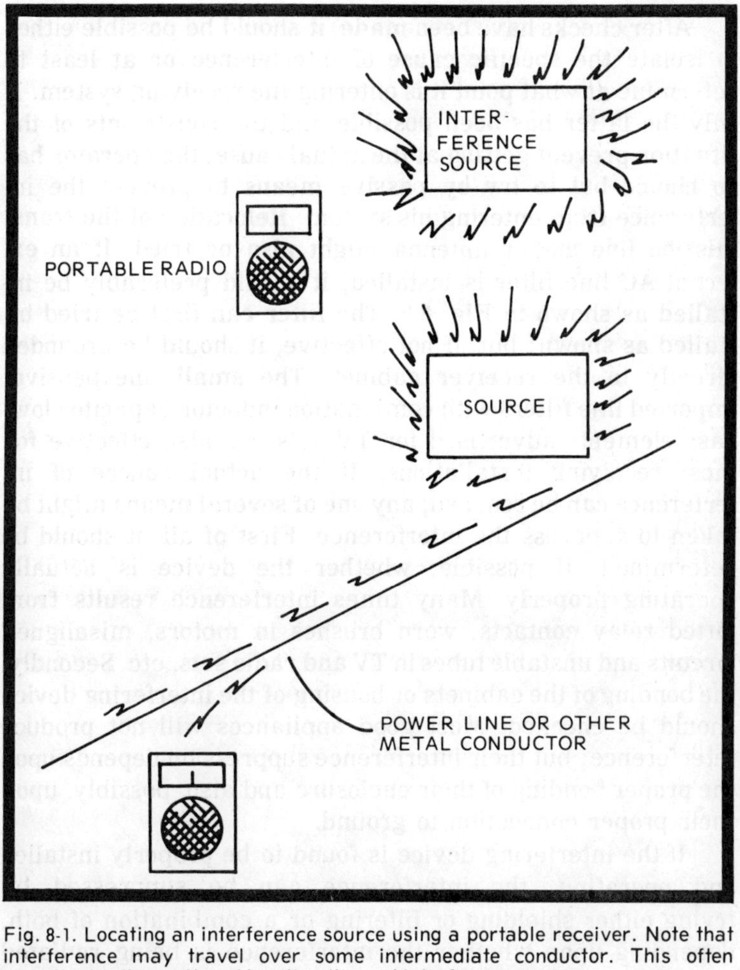

Fig. 8-1. Locating an interference source using a portable receiver. Note that interference may travel over some intermediate conductor. This often complicates the matter of locating the real interference source.

fectly balanced or shielded. However, any increase in noise level several db above that noted when the receiver antenna terminals were directly terminated indicates excessive transmission line noise pickup. Lastly, if no interference pickup has yet been noted, the antenna should be connected. Obviously, then, the interference is entering the receiving system directly from the antenna. One source of interference in congested living areas is harmonic or other spurious signals from older or defective TV or FM radio sets.

SUPPRESSING INTERFERENCE

After checks have been made, it should be possible either to isolate the specific cause of interference or at least to determine at what point it is entering the receiving system. If only the latter has been possible and the constraints of the situation prevent getting at the actual cause, the operator has no choice but to try by passive means to prevent the interference from entering his system. Relocation of the transmission line and-or antenna might also be tried. If an external AC line filter is installed, it should preferably be installed as shown in Fig. 8-2. The filter can first be tried installed as shown; but, if not effective, it should be grounded directly to the receiver cabinet. The small, inexpensive, imported line filters with combination inductor-capacitor low-pass elements advertised for TV sets are also effective for most receiving installations. If the actual source of interference can be reached, any one of several means might be taken to suppress the interference. First of all, it should be determined, if possible, whether the device is actually operating properly. Many times interference results from pitted relay contacts, worn brushes in motors, misaligned circuits and unstable tubes in TV and radio sets, etc. Secondly, the bonding of the cabinets or housing of the interfering device should be checked. Most good appliances will not produce interference; but their interference suppression depends upon the proper bonding of their enclosure and also, possibly, upon their proper connection to ground.

If the interfering device is found to be properly installed and operating, the interference can be suppressed by trying either shielding or filtering or a combination of both, depending upon whether the interference is being radiated from the device, is traveling out via the external leads of the device, or is doing both. Shielding may be achieved by placing some form of aluminum or copper enclosure over the device. This is then bonded to the frame of the device and to a short direct ground lead. In other cases, where a regular enclosure is too difficult to be fabricated, or for shielding a cable run which is radiating interference, a copper foil adhesive tape may prove effective: Scotch type X1181, for instance.

For suppressing radiated interference from spark type contacts, a combination of a 10- to 100-ohm resistor in series with a capacitor of .01 to 1 mfd should be tried. If the contacts are those of a relay or other circuit where the waveshape cannot be disturbed (such as a teleprinter system), higher values of resistance to 2,000 ohms and low capacitance values should be used.

When line filtering is necessary or perhaps the only approach possible, a .01 mfd disc capacitor can be tried from each side of the power line to a short direct ground or, as a less effective connection, to the frame of the device. In simple cases, such capacitors will suffice; but, normally, because of lead and internal inductance, they are not effective for interference above a few MHz. Feed-through capacitors of about .05 mfd are much more effective but more difficult to mount in many small devices. Another AC line filtering possibility is the use of commercial line filters which contain both inductive and capacitive elements; those containing capacitive

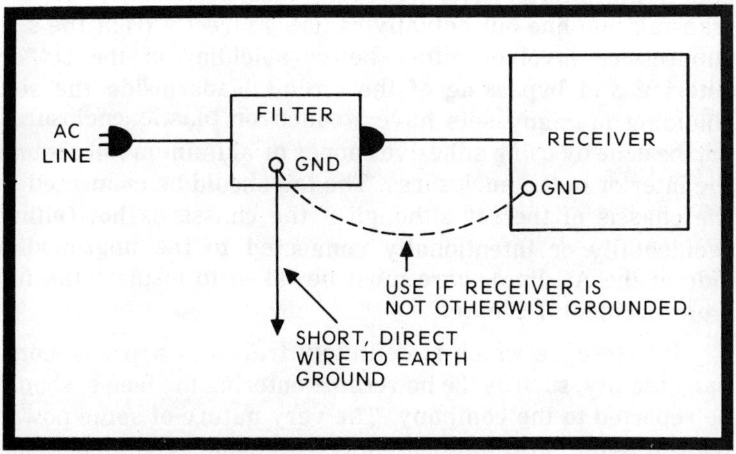

Fig. 8-2. Installation of a power line filter. The ground connections are important and should not be omitted.

elements alone are generally ineffective. Proper use of such a filter, however, requires that it be mounted in the same manner as when used as a receiver line filter.

If a device produces interference only over a narrow band of frequencies, a tuned type of filter can be tried in the AC power line. Such a filter can consist of series tuned inductor-capacitor circuits from each side of the power line to ground or a parallel circuit in series with each power lead. Consideration must be given to making the inductor wire size large enough in the parallel tuned filter so that it does not cause excessive voltage drop to the device. The capacitors used for such filters or even used individually should have a voltage rating of twice the peak voltage in the circuit (a minimum of about 400 volts for 110v lines—and preferably 600 to 1,000 volts when only the DC voltage rating of a capacitor is known).

In the case of a TV or FM radio set which is radiating HF interference through its antenna connection, the easiest cure is a high-pass filter installed in the antenna transmission line and grounded either to a direct ground or, if that is not possible, to the receiver metal enclosure (or chassis, providing it is not **hot** with respect to the AC line). Such filters, available for less than a dollar, are primarily sold to prevent HF interference from entering a TV or FM set (rejecting all inputs below about 40 MHz) but are just as effective in suppressing the interference output in the HF range of a TV or FM set. If the interference, however, does not radiate via the transmission line but actually radiates directly from the set, suppression involves either better shielding of the set or filtering and bypassing of the circuit leads inside the set. Shielding of many sets have wooden or plastic enclosures can be done by using adhesive copper or aluminum foil around the interior of the enclosures. The foil should be connected to the chassis of the set although if the chassis is **hot** (either accidentally or intentionally connected to the ungrounded side of the AC line), care must be taken to insulate the foil from accidental contact.

Interference which is definitely traced to a power company facility, such as the power line entering the house, should be reported to the company. The very nature of some power transmission installations may preclude complete suppression of interference, but generally the power companies have people especially trained in interference suppression and will cooperate to eliminate as much interference as possible. The same is true of the telephone company if interference originates from the dialing or ringing circuits of a telephone installation.

The elimination of interference created by simultaneous operation of a transmitter in the same location as the receiver installation requires first that it be determined if the interference originates in the transmitter itself. The transmitter should be operated into a dummy load connected directly to the transmitter output. A receiver or wavemeter may be used to check that the transmitter output is free of harmonics or other spurious frequencies. Interference which then exists only with the antenna system connected to the transmitter is caused by a rectification process taking place somewhere in the system. Poor metal contacts are the usual culprits. More secure bonding of contacts and the use of wire jumpers across such devices as turnbuckles and other mechanical metal joints are necessary. One must exercise care that interference noted is not due to receiver overload by the transmitted signal. This can be checked by using an attenuator to reduce the input level to the receiver to a low level. With sufficient protection of the receiver against overload by shielding and good tuned input circuits, it is normally possible to operate a transmitter within 10 percent of a receiver frequency without interference.

The elimination of interference in a vehicular installation requires various special techniques which can best be learned from a book on mobile radio installation. In general, the use of spark plug and distributor suppressors, shielding of ignition wiring, a generator filter, voltage regulator contact spark suppressor, static collectors on the wheels, and the secure electrical bonding of all metal parts of the vehicle are the main measures used.

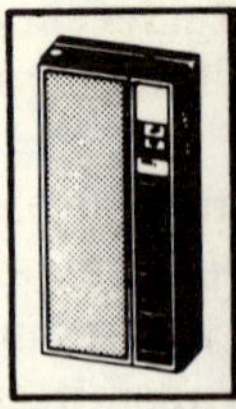

Chapter 9 Professional Receiving—Monitoring Installations

Both commercial and government interests have lead to the establishment of receiving and monitoring stations of considerable size and complexity. Most are stationary in various parts of the world while others, such as military ones used for communications interceptions, are highly mobile. Commercial stations, at least in the United States, are usually associated with companies providing overseas telephone and telegraph service and with maritime communications. In other countries, many news agencies, large newspapers and international broadcasting stations maintain extensive receiving facilities. Government and military stations are concerned with regulatory monitoring, scientific research, command and administrative communications systems and with radio intelligence activities. These stations may range all the way from simple single-operator affairs to installations with large receiver-operator staffs, and even to completely automated complexes where there are no operators, as such, but only equipment maintenance personnel.

TYPES OF MONITORING

The terms **content** and **technical** monitoring are frequently used to distinguish the primary interest of a receiving station. For instance, a news agency station listening to a foreign broadcast is usually only interested in the material broadcast and has no interest in the technical quality of the broadcast beyond that it is intelligible. The same would be true for a user of maritime, aeronautical, public safety or industrial type radio. The interest of many hobbyists also extends only to this aspect of reception. Technical monitoring applies where the signal characteristics are generally of more interest than the material being transmitted. A station engaged in checking the frequency, bandwidth, and modulation characteristics of a transmission would be one example. Another would be the making of an occupancy or time usage chart of a band of frequencies to determine whether excessive frequency space was alloted for some service.

Many monitoring operations are a combination of both content and technical interest. For instance, in a military operation, the initial interest in a transmission may be purely technical until it is learned how to demodulate it, and then the interest will shift to being purely content. Agencies regulating civilian communications generally have both an interest in technical monitoring to see that frequencies and modulation levels are maintained within allowed limits and in content monitoring to see that a station identifies itself properly and does not transmit traffic for which it is not licensed.

The difference between monitoring and receiving gets involved in semantics, but generally a receiving function indicates that the time of transmission is established and that the transmission is intended for the station that listens to it. Monitoring would imply that at least one of these conditions is not met although the terms are often used interchangeably.

SELECTING A STATION SITE

The location of a receiving station is generally a compromise of many factors which may broadly be described as technical and logistical. Technical considerations dictate, of course, that the station be situated in an area where it will be capable of receiving signals from the desired transmitting station. If, for instance, an estimate can be made at what angle signals will arrive, a contour map can be used to determine whether there are any obstructions in the path of the incoming signal. If there are, their effect can perhaps be minimized by increasing the antenna elevation. For long distance reception, effective heights of a wavelength or more are desirable for horizontal antennas; for short distance reception of several hundred miles, effective heights of a half wavelength or less are preferred. If a complex terrain situation exists, such as in a hilly area, such a number of signal reflections are possible that the only sure way to determine the best site may be ground survey with portable receiving equipment.

Receiving stations are best located in areas where the ground is flat or sloping downward for as long a distance as possible in the direction of desired reception. When wooded areas are nearby, they are generally avoided or at least maximum separation is planned between the antenna and wooded area in the desired direction of reception.

Ground conductivity for shortwave reception is not a significant factor, and stations will be found located over varying types of terrain. The advantage of locating near a highly conductive surface, such as sea water, is that a signal which does not reach the station directly may scatter upon reaching the sea surface and propagate along the surface (as a vertically polarized wave) and possibly may thereby be received. For transmitting purposes, a seaside location is usually definitely advantageous.

Most stations interested in receiving sky-wave signals use horizontally polarized antennas. There is no clear-cut evidence that horizontally polarized response is superior. Evidence indicates that shortwave signals which have undergone multiple reflections arrive at a distant site with a mixed polarization so that either horizontally or vertically polarized receiving antennas would be equally effective. Most large-size stations use horizontally polarized antennas, because the antenna types which provide good directivity and are not overly difficult to construct (such as rhombics) have this polarization characteristic.

The other important technical consideration in site location is avoidance of noise sources such as power lines, passing road traffic, and industrial activities. The distribution of such noise in a fairly built-up area is not predictable and really can only be determined by a ground survey.

Most station buildings are located as remotely from noise sources as possible, consistent with logistic considerations. When the buildings are restrained to an undesirable area by logistic considerations, the antennas are located as remotely from the building as possible, consistent with transmission line loss and installation cost.

Logistic factors are all those considerations which have to do with the location of any remote activity—building and land cost, utility services, road access, availability of personnel, communications facilities, etc. Many stations have either their own electric power generator plants or at least a standby generator in case commercial power should fail. Another approach used by some stations which are located near two separate power distribution areas is to have a line run from each.

The buildings themselves are usually arranged to accommodate the type of reception required. When constructed specifically for receiving station use, such buildings usually have extensive ground systems to provide for direct grounding of equipment throughout the station. Also, ductways for cabling are included. In fact, multiple ductways may be provided so antenna transmission lines are separated from all other lines. Audio and teleprinter cables are also kept separated to avoid interference problems. The power lines coming into the station are usually brought through an RF filter and run as directly as possible away from the station, buried if possible. If the station generates its own power, diesel motors are used and the generators shielded and filtered.

TYPICAL EQUIPMENT AND ANTENNA ARRANGEMENTS

Although the function of receiving stations may differ, the flow of signal information follows the same pattern—that is, from the antenna, through transmission lines, to a signal distribution network, to the actual receiving equipment, to auxiliary processing equipment, and then to a terminal function. The facility layouts used differ widely in practice because of space constraints and individual preference.

Antenna fields or parks can be relatively small or can be huge affairs covering many acres. The types of antennas which most stations use for LF reception are loops, short verticals and Beverage. For HF reception, a wider range of antennas is in common usage: horizontal dipole, inverted V, Yagi, vertical wide-band, Adcock arrays, rhombic, and log-periodic. The latter two types are the most commonly used for long-distance reception when space permits.

Antennas may be arranged several ways around the receiver building. The circular arrangement is generally used when reception is desired from a number of directions. The rhombics or other types of similar in-line design can be used for space diversity. Also, rhombics or V antennas which are terminated can be reversed in direction by using relays to set up which end is terminated and which is connected to a

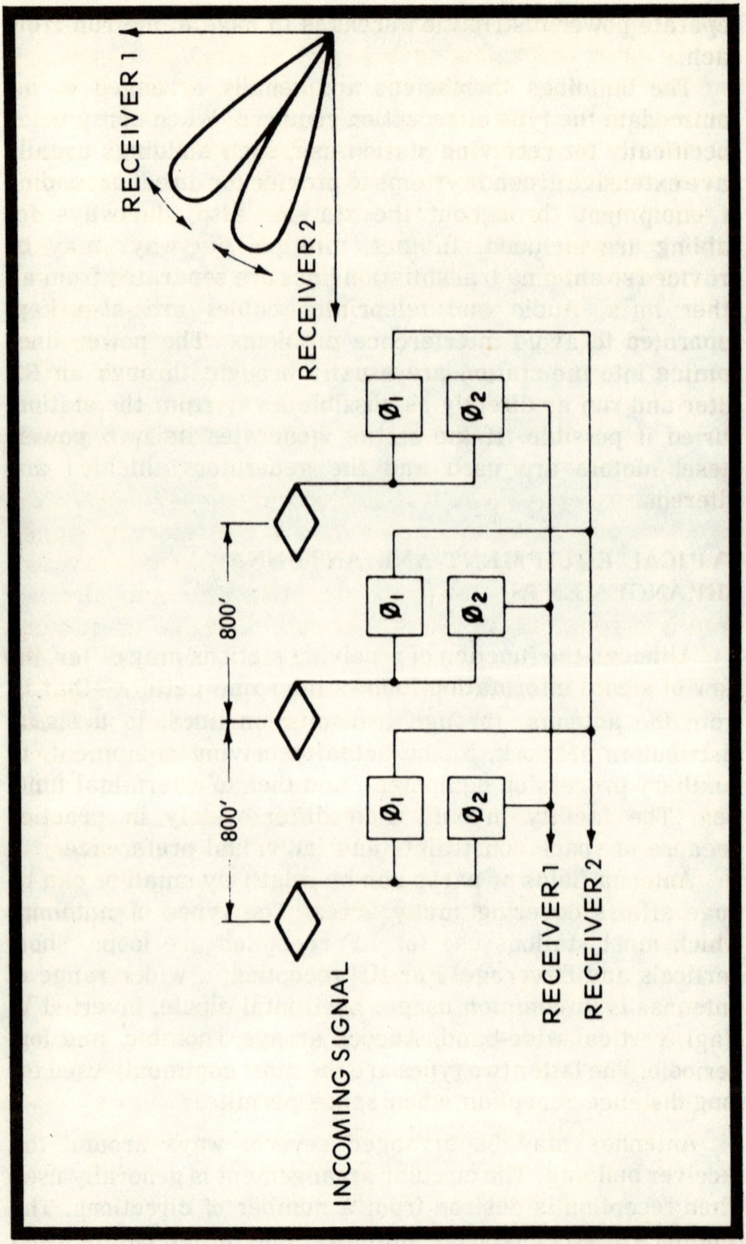

Fig. 9-1. One of the classic large size receiving arrays used for overseas telephone service on shortwave. Three or more rhombics are used and phased together so two or more receivers can receive the incoming signal at different vertical angles simultaneously.

transmission line. If space is limited, a similar arrangement can be used employing inverted-V, terminated long-wires, vertical, or other smaller-space antennas. A triangular or rectangular arrangement of antennas is generally used if reception is concentrated in one direction. Rhombic antennas are also used in other forms such as stacked and overlapping arrays. One of the most interesting uses of rhombics in very large receiving installations is the multiple-unit steerable array shown in Fig. 9-1. The output of each antenna is fed to a series of phase shifters and amplifiers, one group used for each receiver. By adjustment of the phase relationships, the angle of response of the entire antenna system can be varied independently for each receiver. Under conditions where a fading signal arrives at different angles, diversity reception can be achieved by having each receiver output set for a different response angle. It is also possible to have a receiver output automatically track the angle of the incoming signal. The basic system was developed in 1937 and used very successfully for overseas telephone reception. The same idea has been incorporated in other antenna designs to provide multiple sharp, independently steerable outputs in the horizontal or vertical planes from a stationary antenna array.

The separation between rhombic or similar antennas should be sufficient to avoid interaction between antennas designed for the same range. Generally, 1000 feet in the direction of reception and 250 feet lateral separation is used when space and economics permit. In practice, antennas designed for different frequency ranges have been placed side-by-side and even concentrically with no serious interaction.

Other auxiliary antennas which may be found in a receiver installation include omni-directional field strength measuring antennas, ionospheric sounder, scatter sounder, and direction finding antennas. Generally these antennas are placed as far away from other structures as possible to avoid pattern distortion. Frequently the receiver equipment for field strength and direction finding purposes is located in a separate hut or building near their associated antennas.

The transmission lines for the other antennas are brought into a central antenna distribution location in the main receiver building. The cables themselves are usually buried both for shielding and physical security. Coaxial cable, often RG-17-U when long runs are involved, or pressurized, semi-

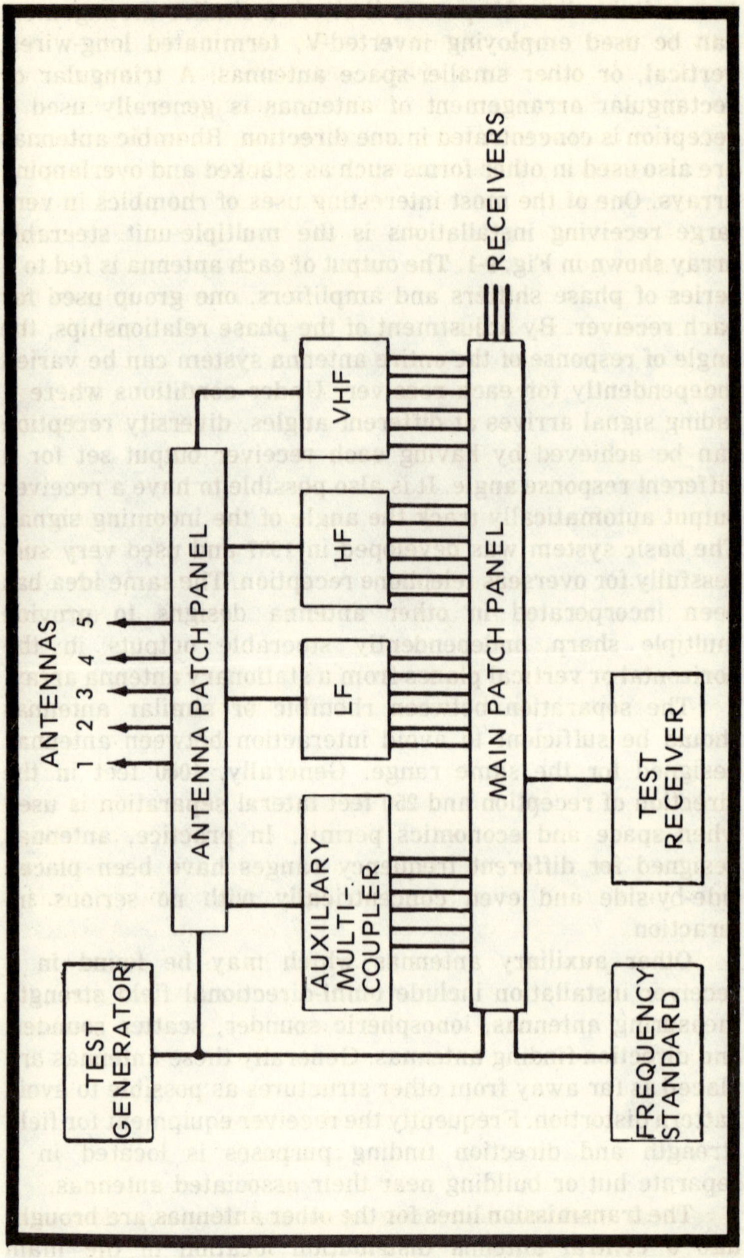

Fig. 9-2. Large size receiving installations often require elaborate interconnection schemes. Frequent use is made of patch panels to allow the flexible interconnection of all facilities.

rigid coaxial cable is often used. Matching transformers, when necessary, are connected at the antenna to match the coaxial cable. Some form of lightning protection, either of the gap type or heavy-duty RF chokes connected from the antenna termination to ground, is also included at each antenna.

The antenna distribution within the building might follow the general scheme shown in Fig. 9-2. The first patch panel would allow usage of the multi-couplers as desired, including substitution for a defective unit. The main patch panel provides for the distribution of the multi-coupler outputs to both the receivers directly and to auxiliary patch panels. A test generator input is provided so periodic maintenance checks of the system can be made. A frequency standard signal may also be distributed from a central standard when it is too expensive to have one at each receiving position.

The arrangement of the receiver equipment itself, as mentioned before, depends upon whether the station's primary interest is in content or technical reception (monitoring). The positions for technical monitoring are set up with a grouping of equipment so that functions such as frequency measurement, modulation characteristics, band usage observations, field strength measurements, and direction-finding can be accomplished from one area. A small antenna patch panel will normally be provided at each receiver location. Many such stations provide a self-regulatory function for their associated transmitting installations by checking technical signal characteristics. They also check these characteristics for foreign broadcasters in return for similar information on their organization's transmissions to foreign areas. They also assist in cooperative efforts to monitor the various broadcast bands for identifying sources of interference and checking frequency occupancy. The latter information combined with propagation studies allows their organizations to decide upon the most advantageous transmitting frequencies or to dispute any usage of frequencies which are registered to them.

Some of the methods used to accomplish the above services may be of interest. Frequency measurement can be done by several methods depending upon the accuracy desired. Two relatively simple methods are possible. In one, a frequency synthesizer provides output every one kHz. The nearest kHz signal lower in frequency than the signal to be measured produces a beat note with the incoming signal. The resultant audio beat note is then zero-beated against an audio oscillator using either aural means or an oscilloscope display. The received frequency is the sum of the synthesizer and audio oscillator frequencies. In the other method, a tunable oscillator is used to directly zero-beat the incoming signal. The frequency of this oscillator, and hence that of the received signal, is read directly by a frequency counter, models of which are available to measure within one Hertz frequencies up to the VHF range. These methods work well when sufficient carrier is transmitted such as for AM and CW transmissions but must be modified for other transmission modes.

Modulation characteristics can be studied on a panoramic display unit. Overmodulation and the resulting "splatter" can usually be seen on such a unit, as also may any major frequency instability. However, it is an expensive means to use to make an estimate of the percentage of modulation for an AM signal. This is done easily on a normal oscilloscope by displaying the diode detector output of the IF signal (DC and AC components). With 100 percent modulation the display will vary between zero and twice the value produced by the unmodulated carrier. A still more interesting display can be made by also connecting the audio signal from the receiver to the horizontal input of the oscilloscope for use as the sweep voltage. Different size coupling capacitors may be necessary to correct for phase differences so that the pattern will be flat. The flat triangular pattern indicates 100 percent modulation when all corners are closed with no extending tips. The straightness of the sides of the pattern indicates the linearity of the modulation. The percentage of modulation is of interest to most broadcasting stations because overmodulation may cause distortion and violate broadcasting regulations. On the other hand, a low percentage of modulation means full use is not being made of a transmitter's capabilities. Receiving stations can also measure other modulation characteristics such as harmonic distortion with panoramic units, but this can only be done accurately with the transmitter sending various test tones.

Field strength is of interest because regulations, especially for domestic stations, may limit it to prevent interference between stations on the same frequency. It may also be used to evaluate the results of improvements in transmission facilities. Field strength (VU meter reading) is the product of the voltage induced in an antenna times its effective length. For a dipole effective length is only .64 of the physical length. Of the voltage induced in the antenna, half appears at the receiver terminals. The voltage at the receiver terminals can be measured directly if a receiver with calibrated gain is used or an ordinary receiver can be used with an accurately calibrated signal generator. The receiver input is switched to the generator which is adjusted to give the same output level from the receiver as the received signal produces. Field strength can then be calculated. In practice, most stations use standardized field strength antennas and calibrated receivers so that making such measurements is a relatively simple procedure.

Some stations monitor certain portions of the HF bands regularly and carefully catalog the stations heard. The results of such observations are summarized by regional or international organizations so some over-all compilation of broadcasting activity is achieved. Other stations only monitor various frequencies if reports of interference are received. The determination of the interfering stations is made aurally (or with print-out equipment for teleprinter transmissions) sometimes with the assistance of direction-finding receivers. Paper-ink recorders are used for frequency occupancy studies. The recorder mechanism mechanically tunes a receiver across a specific frequency segment and makes a pen recording of the received signal amplitude versus frequency. The receiver selectivity is usually set at one kHz or less for good resolution and a sweep made about every 30 seconds. If enough sweeps are made, a frequency versus time (one hour, for instance) record is obtained which can be studied to aid in study of propagation conditions, interference patterns, etc. Aural identification must be made of the recorded signals, however. Synchronized tape recorders are also used sometimes. Similar recorders or ink pen types with much slower sweep rates are also used to produce a record of the received field strength of a specific station.

Often monitoring stations cannot be manned physically at all the hours when interesting reception may occur. In this case, frequent use is made of clock timers or clock switches such as that shown in Fig. 9-3 to turn on a receiver and a tape recorder for reception of a station at odd hours. Such timers are available in a wide variety of styles.

Fig. 9-3. Frequent use is made of clock switches to turn on receivers, recorders, etc. when an operator is not present. Such units are available for daily switching at specific times and can even be adjusted to skip days of the week.

DISTRIBUTION AND PROCESSING

The output of content and technical reception stations differs, of course, according to their function. The output of a content type operation is the received program material as such. These programs are usually routed to a patch panel and then to the means of transfer between the receiving station and the using activity. The using activity may be a transmitter site in cases where programs are re-broadcast or relayed further by radio. A good example of such activity is the distribution of Voice of America programs produced in Washington, D.C. These programs will be sent to transmitter sites either in Greenville, North Carolina or Dixon, California. From these sites the programs are relayed to various receiver sites in Europe and the Far East for local rebroadcast. The user activity might also be a new office or other editorial group. Journalists will analyze the received programs for their news or information content. They may also make translations of programs when required for further distribution of news items or other material of interest. Although not employed extensively in the United States, such news gathering means is employed in varying degrees by major foreign newspapers, partly because of their location adjacent to areas of news interest and because of the lack of independent news agency services in many countries.

Commonly the receiver site will be connected to the user activity either by land line (telephone or higher-quality circuits) or by VHF-UHF multiple channel radio relay. Which means is used depends, of course, upon the distance between activities and the relative cost, reliability and availability of the facilities. The number of channels between activities might vary anywhere from one to over a hundred. The output of a technical receiving station, on the other hand, most often does not leave the station in electrical form. The recorded material and the written reception or measurement reports made out by the operators may still receive further processing, but the received signals usually terminate at the station. Government stations monitoring for regulatory compliance might contact an offending station immediately. However, most technical receiving stations will issue only periodic reports—weekly or monthly—on their observations of band occupancy, field strength measurements, frequency measurements, etc.

REPORT FORMS AND LOGS

Many forms of reports, logs, and schedules are used by receiving stations. A station engaged in handling scheduled overseas transmissions regularly every day might use a bulletin-board type of schedule. Another possibility is the use of a loose-leaf notebook with a separate page, in time order, for each station which gives, besides the time of reception, the receiver to be used, antenna, main and alternate frequencies, call sign, sign-on or program type and recording or switching instructions for distribution of the program.

Reception logs differ according to the type of reception. A content reception station might only use a log to indicate if a station was generally received satisfactorily according to the reception schedule. Only if it could not be received satisfactorily might a detailed note be made as to whether lack of reception was due to weak signals, disturbances, interference, equipment breakdown, etc.

A technical monitoring station will usually make out a more detailed reception report especially if engaged with another broadcasting organization in mutual checking of transmissions or if preparing a reception listing for publication. The format used for such reports usually follows the SINPO code.

Signal characteristics may be rated on a scale of one (practically none or extremely poor) to five (full or excellent quality). Letters are often used also to clarify the number ratings. More elaborate forms might clarify more items or even specify what the number scales mean for each item. (For instance, a three for modulation quality might mean 40 to 60 percent modulated; or a five for signal strength might mean one millivolt per meter or more.) There are no international standards for such ratings, so those used are often specified on the reception reports; sometimes such reports will also be accompanied by tape recordings to further clarify reception conditions.

Another type of reception report that might be made is on some standard station such WWV to check propagation conditions. Only the strength of the station will be recorded periodically on several frequencies in order to give an indication of changing propagation conditions for various parts of the shortwave spectrum. The propagation forecasts sent by WWV might also be recorded at the same time as the signal strength in order to provide another means of correlating general reception possibilities to propagation conditions. Still another reception report is that of the sunspot counts for each day as broadcast by Radio Schwarzenburg, Switzerland. A station might use this information as well as other ionospheric measurement data to issue their own propagation forecasts, concentrated on their organization's frequency and broadcasting area interests.

A form of reception report is also made from band occupancy recordings. Either the charts themselves can be used or an hourly summary (using a one to five scale for clear channel to completely covered channel) can be made for specific frequencies of interest so that an overall evaluation of the clearest frequencies available at specific times can be formed.

Card files are also used extensively as a means of recording reception results. If a study is being made of the stations in a certain international broadcasting band, a card would be filled out for each station and notations made regarding program schedules, field strength, modulation quality or on whatever other characteristics were of interest.

Other forms used in a station would concern the maintenance of equipment, to insure that performance does not deteriorate. The frequency standard might be checked and logged daily using the transmission from some standard station as a reference. If several frequency standards are available, a continuous check of one against the other might be made by means of a pen recorder to indicate any abnormal change in one standard. Field strength receivers might be checked weekly for accuracy while receivers used for normal reception might only be checked monthly for sensitivity, bandwidth, frequency calibration, etc. Antenna system cables, switching facilities, and audio and teleprinter distribution systems usually are also checked for distortion, loss, and crosstalk on a monthly basis. Careful records can often highlight a deteriorating facility or circuit before an actual breakdown occurs. There are no set forms for these records, although most stations use a card file system with one card used for each piece of equipment, transmission line, or antenna.

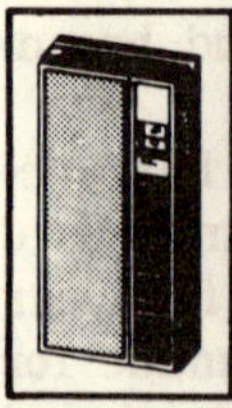

Chapter 10 Testing and Maintaining the Receiver Installation

The care and repair of electronic equipment can be categorized under two terms: **preventive** and **corrective** maintenance. Preventive maintenance means caring for equipment properly so that it does not become defective, or at least noting as soon as possible when a defect does develop, and corrective maintenance means to do something about the defect. A hobbyist can easily do most of the work required by the first category and a surprising amount of the second type.

By far the great majority of service problems that equipment owners refer to professional technicians involve minor repairs which, with some patience and common sense, the owner could have discovered himself. A smaller but still significant number of service problems are caused by equipment owners who try to adjust and repair equipment when they do not have any clear indication of what they are adjusting or whether their procedure can really correct the defect. They hope that by adjusting some control a defect will suddenly be cured. Hopefully, the material in this chapter will enable the reader to judge what he can do, himself, and when he should seek professional service for his equipment.

A purpose also of this chapter is to describe the function and use of test equipment used to measure the most important performance characteristics. A number of standard tests and troubleshooting routines are described which can be performed with a minimum amount of digging into a receiver. Most receiver instruction books provide simple suggestions for troubleshooting and should be carefully read since supposedly they were developed with the characteristics of the specific receiver in mind. Few such books, however, provide information on exactly how the performance characteristics claimed for the receiver were measured.

Even if the reader does not yet have or even immediately contemplate buying any test equipment, it will do no harm to be familiar with the basic receiver test routines. A number of the receiver performance standards mentioned earlier really only have meaning when one understands how they are measured.

SIMPLE RADIO RECEIVER CHECKS

Among the main receiver checks a hobbyist can periodically make are sensitivity (ability to pick up weak stations), tuning accuracy, the frequency response, audio distortion, and the power output of a built-in audio amplifier. All these things can be checked easily if you have access to the proper test equipment. However, it is assumed to start that this is not the case and that you merely want to periodically check the performance to detect the development of any major fault.

The best way to check receiver sensitivity is to observe whether or not distant stations can still be received. However, the reception of distant stations varies greatly from time to time; therefore, such a sensitivity check must be a continual process, rather than a test made at a specific time. What you need is a constant signal source which will provide a known reading on the receiver signal-strength meter.

A good, constant signal source is a local or nearby station. When the receiver is known to be working properly, you should record the S-meter reading. Then at any time you can tune in that station and check the present reading against the standard. If the two agree, you know the receiver sensitivity is normal. Of course, the same antenna must be used if a comparison is to be valid. A more sensitive test can be made with the antenna disconnected from the receiver in each case. If there is no reception with the antenna disconnected, you can try adding a small wire antenna of minimum length (a few inches) until only the strongest station is heard. This "calibration" antenna is used in the future to see whether or not the same reception is still possible. In essence, you are duplicating the reception of a very weak station by this method, yet using a station which should always have approximately the same signal availability. If the receiver does not have a signal-strength meter, you can still use this check, basing your judgment on a barely audible sound level at a reference volume control setting.

The receiver's tuning scale accuracy can be checked by comparing the scale reading for a station to that published in station schedules, program guides, etc. The check is useful mainly to indicate that the oscillator portion of the receiver is stable and that the mechanical portion of the tuning mechanism is not becoming loose. If the point at which different stations are received keeps changing, it could indicate that the oscillator section is developing problems or there is slippage or backlash in the mechanical tuning mechanism.

The frequency response of the entire receiver can most easily be checked by finding a strong station which periodically sends a test transmission of different frequency tones. Such transmissions are usually broadcast at the start of special hi-fi programs. These checks are only useful, however, when coupled with a check of the audio output level of the receiver at a reference setting of the volume and tone controls. That is, if you are receiving the station at the same signal level as usual and the receiver controls are all set the same as before, the audio output level as measured by a VU meter across the speaker line should be the same for each audio tone as it was during the first check. The level normally will not be exactly the same for each specific audio tone.

The power output of the audio section of a receiver or amplifier in absolute terms is near impossible to measure without the use of proper instruments. This difficulty arises because power output alone does not have much meaning unless it is coupled with some knowledge of the distortion produced. In simpler words, it is not only the quantity of audio produced that is important but its quality as well. You can obtain an indication of whether the power output is falling off or not, however, during the frequency response test just described. You can also play a tone test tape through the audio section of the receiver and notice if the output level remains the same for each tone frequency at the reference settings of the amplifier controls on the receiver.

Distortion can be detected by a listening test only in a very coarse sense because distortion of several percent may not be noticeable and the ability to hear distortion varies from person to person. The best that you probably can do is to simply listen to the tones on a test tape at both very low and loud volume settings and try to detect any build-up of distortion. If you can turn the volume up enough (or by using headphones), there will occur a point where distortion becomes apparent in any case. Any change in this point to a lower volume setting could indicate that some problem is developing within the system.

The above checks are by no means meant to measure anything in specific terms. They are, in fact, only a possible, orderly way to check by ear whether any condition in a receiver is changing. The important point is that any shift in a receiver's operating characteristics when it is used consistently under exactly the same condition indicates that some change is taking place in the receiver circuitry which could mean that a problem is developing.

CLEANING THE CONTROLS

One of the most common complaints which develops in any receiver or amplifier after a period of time is noisy controls or switches. Such problems are due to the fact that these units cannot be sealed completely from the atmosphere at a reasonable cost. Therefore, with time, dust will find its way into switch contacts and potentiometer surfaces.

It is a relatively simple matter to clean such controls and switches, thanks to the aerosol spray can, and it should be done on a regular basis. There are many types of control cleaners on the market, variously labeled as contact cleaners, control cleaners, tuner cleaners, etc. Any of these sprays can be used interchangeably. Be sure that the spray is labeled as being safe for plastics and that it does not leave a residue. A spray can that comes with a plastic extension tip from the nozzle is very convenient when you have to reach into difficult corners.

The use of a spray is simple. The idea is to spray the cleaner as near as possible to or on the contacts that are to be cleaned. In case of a front-end which is bandswitched, this may mean that the cover of the set may have to be removed to get at the contacts. But before you go to that trouble, try to remove the tuning knob and spray in from the front of the set. Volume controls and other potentiometer-type controls should be sprayed from the front at the point where the rotatable shaft enters the control; the knob should be removed from the shaft, of course. Since the spray cleaner only has the effect of washing dirt away and is quite mild in its action, the switch or control being cleaned must be rotated or switched slowly while the cleaner is being applied; otherwise, the cleaner will not be very effective.

If you have difficulty reaching certain controls with a spray cleaner, or if there is a problem about getting the spray on the front panel of the equipment, adapters are available which fit over the control shaft (once the knob has been removed) and screw onto the threads of the potentiometer body. A spray burst directed down the hollow center of the adapter is guided into the potentiometer. If such an adapter is not readily available, any hollow plastic tube can be used for the same effect. The tube must, however, be wider than the potentiometer shaft, since the cleaner must enter the potentiometer where the shaft rotates in the threaded portion of the potentiometer case.

VISUAL INSPECTION

Many faults and potential troubles can be found by a simple visual inspection of the item of electronic equipment. If it fails to function when switched on, or if an obvious fault is noticed (lack of power, no sound, distortion, etc.), you should switch the equipment off again and conduct a careful visual inspection of the components. If the equipment fails to turn on at all, a burnt out fuse may be all that is involved. Obviously, if new fuses continue to blow, a more indirect fault may exist. Carefully check the components to see if any is clearly burned or broken. Small resistors often break in two. Faulty capacitors may develop bubbles on their otherwise smooth surfaces. A burnt out power transformer often leaves a residue of wax and resin by the base.

In a tube-type set, you can search for faulty tubes by noting if their filaments light. For some receivers, this means that you must arrange to bypass the power disconnect on the back of the set, since when the back is removed the power is disconnected. Special "cheater" power cords, such as are used by TV-servicemen for this purpose, are available. With some tube types, the lighted filament is difficult to see and a check may better be made by touching the glass bulb to see if it is warm. Difficult-to-see corners can be reached by means of a pencil-type flashlight and a small pocket mirror.

A visual inspection can be concentrated in a given section of the equipment if the fault can be isolated to a specific function. For instance, if the S-meter indicates reception but no sound is heard, the search should be concentrated around those areas designated on the tube chart as having to do with sound output. If a radio receiver's audio section works fine on tape playback but no station can be received, suspect trouble in the front-end, IF sections, etc.

Visual inspection is greatly expedited if you have some basic knowledge of which sections perform various basic functions. Even lacking this knowledge, such an inspection is worthwhile if done with patience. Probably half of the faults in a piece of equipment are revealed by some form of visible indication. If a clearly defective component is located, it can be replaced after it has been properly identified.

CHECKING CONNECTIONS

Frequent causes of trouble in receiving installations are the cable and component connections within the system. An ohmmeter is very useful to check the continuity of cables and to assure that short circuits do not exist. Where wires enter connectors, there is usually strain, and such points should be among the first checked. Also, wires that must flex, for example, those in recorders where the recording heads move in and out, are prone to break after extended use. Often such breaks occur in the internal wire strands and are not visible because of the insulation. An ohmmeter must be used to check such leads. Check the leads to switches, also. Many switches mounted on printed-circuit boards appear rigid, but they actually flex during operation.

Soldered component connections usually do not change. However, there are cases where a solder connection has not been properly made to begin with; a so-called "cold" solder joint makes only intermittent contact. Such a joint is difficult to discover because it may appear to be good. If you have isolated such a difficulty to a certain section in a piece of equipment, it is often necessary to resolder every connection. Another approach that sometimes works is to spray a cold-producing liquid on each joint while listening for the result. The freezing of the joint may make the elements of the joint contact separate enough to open up the connection. Special freezing sprays are available, but contact or tuner cleaner can be used, too, since the carbon dioxide propellant in the spray will also cool the joint.

Printed-circuit boards which flex during the operation of a switch or tuning control can develop connection faults particularly difficult to detect. These faults are often hairline cracks in one or more of the metal runs on the board. If the board is small and simple enough, you can also go over the board, carefully running solder over all metal surfaces. Be careful, though, that solder slivers do not inadvertently connect adjacent circuits!

In installations subject to weather conditions, corroded connections are to be suspected. Usually, such connections can be detected visually, but there are instances where connections were not made correctly to begin with and each metal surface has built up an oxide coating. This condition is not visible, but, nonetheless, does open the connection. Flexing cables are another problem source outdoors, especially if the cable ends have not been properly sealed. Moisture can enter the cable and corrode the inner conductors at a point within the cable away from the end. Also, the appearance of corroded spots anywhere along the cable is a sign that moisture has entered the cable, perhaps through a nick in the insulation, and is corroding the inner conductors.

REPLACING COMPONENTS

Once a defective component has been located, the problem is to properly identify it. Resistors are usually standardized and color-coded. Most capacitors are marked with a value, except for some mica types which are also color coded. If it is available, check the wiring schematic of the equipment to see that the value stated on the component removed corresponds to what the schematic calls for. At times, you will find components which are marked only with a manufacturer's part number. The schematic may reveal the component value; or you may just have to make an intelligent guess, depending upon whether it is possible to trace the circuit in which the component is used. In many cases, guesswork can be dangerous, so it would be wise to consult a technician familiar with the equipment.

Generally, you should try to replace a component with one of the same value, unless, for instance, a certain resistor is constantly burning out because its wattage rating is insufficient. When replacing small size capacitors, it is usually best to use the same value. When replacing larger capacitors (5 mfd or larger electrolytics), the exact value is rarely critical. The tolerance on such capacitors as they are normally manufactured may range up to 50 percent away from their nominal value. Except for electrolytic capacitors, as just mentioned, the factors that should be observed are the resistance and capacitance ratings. The wattage rating of a resistor or voltage rating of a capacitor may be any value at least as high as that of the component being replaced.

Diode replacement is usually simpler in the sense that its number specifies its characteristics. Unless a diode is used in a critical application, a general purpose replacement closely matching the specifications of the original can be used. Zener diode replacements depend upon the circuit application. For instance, in a speed control circuit in a recorder, an 8.5-volt type should not be replaced with a 9.1-volt type, although this substitution would probably work fine for regulating the voltage to an audio preamplifier. Zener diodes also have dissipation ratings, similar to resistors, which must be observed. The diodes used in power supplies are frequently unmarked. Since such diodes are relatively inexpensive, and lacking any exact data, you should replace them with types that have a voltage rating of several times that of the DC voltage output of the supply (as marked on a schematic) and a current rating several times that of the power output of the unit divided by the normal DC supply voltage.

Transformer replacements should be made with equivalent types. There are several universal power transformer replacements. For other equipment where the transformer has a single secondary winding and the transformer ratings are available, you should compare the manufacturer's replacement cost against that of a mail order house replacement type. The latter will often be much less expensive. If there is some special feature about the transformer, such as multiple secondary windings, a manufacturer's replacement may have to be used. The only other possibility may be to replace a single transformer with two or more single secondary types, if there's room in the equipment.

CHECKING AND REPLACING TUBES

If all the tube filaments light and you have checked for obvious component connection faults, it may still be worthwhile to check the tubes in a tube tester. Many faults occur within a tube besides a filament burn-out. On the other hand, checks with a simple tube tester do not always guarantee that the tube is really good, although this will be the case 80 to 90

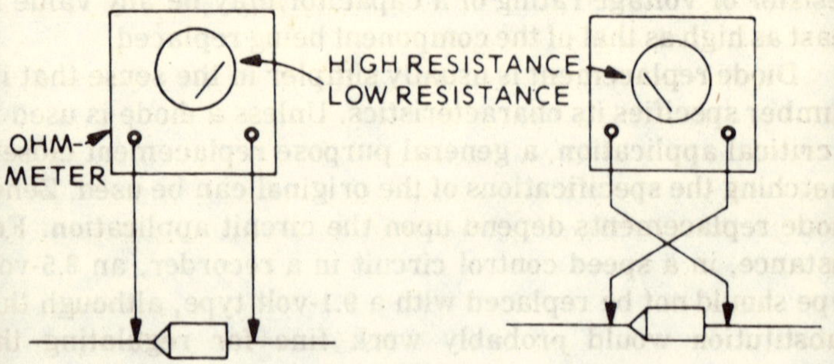

Fig. 10-1. Testing a diode rectifier using a simple ohmmeter.

percent of the time. Some exceptions develop because no tube tester can completely duplicate the actual circuit conditions under which a tube is used and because some testers check only the tube as a diode and not as an amplifying device.

Many tube types are interchangeable in general purpose applications. Tube substitution manuals are available, and stores that sell replacement tubes usually have cross reference lists. Even if you do not use the mail order houses for securing tubes needed in a rush, it is still a good idea to use their price lists as a reference to check what you are being charged locally. Unbranded tubes are usually somewhat less expensive than tubes from major U.S. manufacturers. Generally, there is no difference in performance between branded and unbranded types in normal receiver applications.

CHECKING AND REPLACING DIODES AND TRANSISTORS

Diodes for most applications are easily checked with an ohmmeter, as shown in Fig. 10-1. The ohmmeter leads are placed across the diode terminals, the meter reading noted, and then the ohmmeter leads reversed across the diode terminals. You should read a high resistance of at least several thousand ohms in one direction and a low resistance of less than a hundred ohms in the other direction. As long as the difference is in a ratio of 10 to 1 or more, the diode is probably good. If some doubt exists, try to test the diode in the manner described and then check the results against those obtained with a diode known to be good.

Soldering in most power type diodes does not require any special precautions, although the soldering iron should not be applied any longer than necessary. When installing small signal diodes, a heat dissipator should be used or a small loop should be formed from the end lead on the diode before it is soldered in place.

Transistors can be checked in basically the same manner as diodes by simply treating them as two separate diodes with the base as the common lead. This technique is illustrated in Fig. 10-2. The readings for each "diode" will not be the same, but each diode should act as a good diode. This test technique works fine for the great majority of transistors, which are of the junction type. Field-effect transistors cannot be tested in this simple manner. Commercially made transistor testers make more exact tests on the gain and other characteristics.

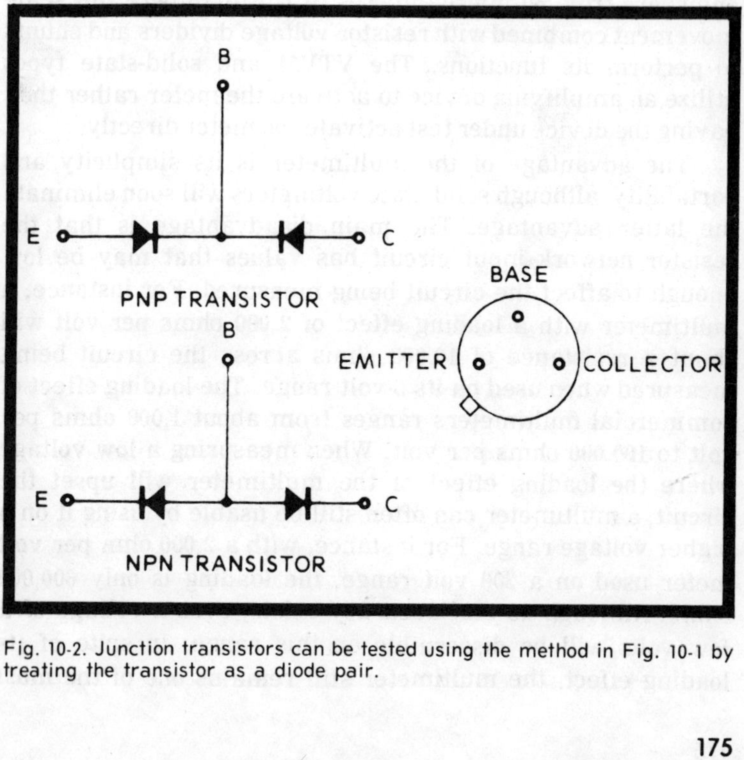

Fig. 10-2. Junction transistors can be tested using the method in Fig. 10-1 by treating the transistor as a diode pair.

Since transistors do not age in the same manner as vacuum tubes, however, there is no need to periodically check the state of transistors.

Transistors are replaced by simply soldering them in place, again with a minimum application of heat. If you are not practiced in soldering, a heat dissipator should be used, or better yet a transistor socket. The latter is soldered in place, then the transistor leads are clipped and pushed into the socket. This technique is especially recommended for the replacement of field-effect transistors where static electricity effects during the soldering process can ruin some types of these transistors.

BASIC TEST EQUIPMENT

The most basic and useful piece of measuring equipment is the multimeter and the vacuum tube voltmeter (VTVM) or solid-state voltmeter. Any one of these instruments may have scales calibrated in terms of AC and DC volts, AC and DC current, ohms and decibels. The multimeter uses a meter of anywhere from 50 microamperes to 1 milliampere full scale movement combined with resistor voltage dividers and shunts to perform its functions. The VTVM and solid-state types utilize an amplifying device to activate the meter rather than having the device under test activate the meter directly.

The advantage of the multimeter is its simplicity and portability, although solid-state voltmeters will soon eliminate the latter advantage. The main disadvantage is that the resistor network input circuit has values that may be low enough to affect the circuit being measured. For instance, a multimeter with a loading effect of 2,000 ohms per volt will shunt a resistance of 10,000 ohms across the circuit being measured when used on its 5-volt range. The loading effect of commercial multimeters ranges from about 1,000 ohms per volt to 100,000 ohms per volt. When measuring a low voltage where the loading effect of the multimeter will upset the circuit, a multimeter can often still be usable by using it on a higher voltage range. For instance, with a 2,000 ohm per volt meter used on a 300 volt range, the loading is only 600,000 ohms. Although no real accuracy exists, even a voltage of a few volts will be discernible on this range. In spite of its loading effect, the multimeter still remains one of the most versatile intruments for measuring voltages in power supplies and other low-impedance circuits.

The main advantage of the VTVM or solid-state equivalent is the very slight loading effect it has — from 1 to 50 megohms. It can be also made to measure much lower voltages than a multimeter (down to microvolts or less) and has a wider frequency range for AC voltage measurements.

Two precautions should be noted with either general type of instrument. When measuring ohms, both instruments pass some current from an internal voltage source, usually a battery, through the resistance under measurement. The lower the ohmic range on the instrument, the greater the current passed through an external resistance. This factor should be kept in mind when using an ohmmeter function to test solid-state devices. The other precaution to note is that the AC voltage scale on a meter is normally calibrated in terms of RMS voltage. The reading is only correct for a pure sine wave voltage measurement.

Most of the scales on a meter are self explanatory—except for the db scale. The decibel is simply a means for expressing a ratio. Instead of saying a voltage, current, or power is, for instance, 1000 times greater than another voltage, current, or power, one can say that it is 60 db greater (for voltage or current) or 30 db greater (for power). The difference in db when considering power versus voltage or current arises because the decibel, when developed for use in telephone work, was defined as a power comparison unit. Therefore, since power varies as the square of voltage or current, the latter two in db are always twice the db value of the same numerical power ratio. An example of a modern solid-state multimeter is shown in Fig. 10-3.

An RF signal generator is one of the next most basic instruments used for a great deal of receiver testing. Essentially, it consists of a variable frequency oscillator which in several ranges will cover from the lowest IF frequencies used (usually 50-1000 kHz) to the highest signal frequency at which a receiver operates. One of the most important features of such a generator is that it has a calibrated output — usually ranging from essentially zero to 1 volt or more and continuously adjustable over this range. The output level may be read from a meter, or an output control may be used to adjust the output to some set level using a meter or other indicator. The calibration marks on an output attenuator then indicate

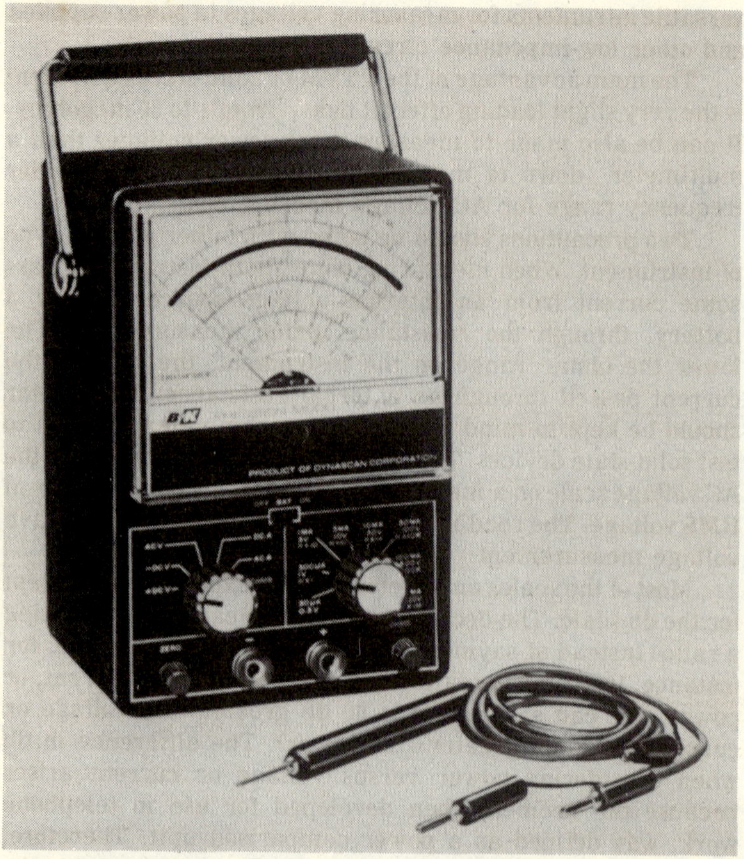

Fig. 10-3. One of the first investments to make in test equipment is a VOM or preferably a good solid-state multimeter such as that shown.

the output in microvolts or millivolts. Some inexpensive generators have a variable but non-calibrated output, and the lack of this feature precludes their usefulness for receiver sensitivity measurements unless one also has available a rather expensive AC VTVM to measure the very low generator output voltage levels. Frequency accuracy varies widely depending upon the quality of the generator. An extremely accurate readout is not necessary for general sensitivity, gain, and bandwidth measurements. However, for testing such things as exact IF selectivity characteristics, both a high degree of stability and exact frequency calibration are important. A signal generator also normally has some provision for modulating the RF output with a 400 or 1000 Hertz tone in order to perform standard AM sensitivity tests.

Marker generators are often built as separate test units, especially if the calibration of a signal generator is not too accurate. For instance, such a generator might provide a crystal controlled output on 455 kHz, to provide for IF alignment of most receivers and 1,000 and 100 kHz outputs with strong harmonic content so that markers are available every 1 MHz and 100 kHz throughout the shortwave range for the calibration of a signal generator, receivers, etc. If a signal generator calibration is being checked, a receiver can be used to detect zero beat. Some marker generators have a mixer stage incorporated so the signal generator output can be fed directly into the unit and headphones may be used to detect zero beat.

Marker generators are sometimes referred to as secondary frequency standards. A primary frequency standard is one containing a highly stabilized crystal oscillator, the output of which is compared continuously to some standard such as WWV transmissions. Secondary standards lack any continuous method of comparison. When a simple crystal oscillator is stable enough to be regarded as a secondary standard is not defined.

Audio generators provide a pure sinusoidal wave output over the frequency range of about a few Hertz to 100,000 Hertz. Some expanded range generators will extend up to 1 MHz. As with an RF signal generator, the usefulness of the instrument depends upon having available a means, either internal or external, to measure the output level. Such generators are useful to measure the gain and response of audio circuits. Many audio generators will also provide for a square wave output capability. The square wave function allows easier testing of audio circuit transient response and phase shift, and as a switching circuit can be used as a drive source for electronic switches. As with an RF generator, the hum and other spurious outputs should not be more than 1 percent of the output level and the waveform distortion should be no more than a few percent if reasonably accurate measurements are to be made. Generally, an RF signal generator will be more useful initially to the receiver operator unless he also has a strong interest in audio circuitry.

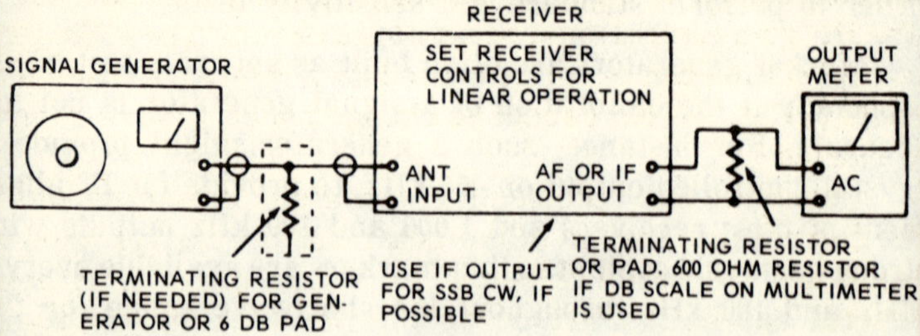

Procedure:

1. Establish the reference output level (generator output off for CW and SSB, on but unmodulated for AM). Peak the output by tuning the generator across the receiver frequency.
2. Turn the generator on for CW-SSB, with 30% modulation for AM. Use the lowest possible generator output to start.
3. The CHANGE in receiver output level in db is the S + N divided by N ratio for a given carrier level at the receiver input.
4. Sensitivity is valid for a specific IF bandwidth only.

Fig. 10-4. Test procedure for a sensitivity check.

PERFORMANCE CHECKS USING TEST EQUIPMENT

The following paragraphs describe a few basic receiver tests which can be performed with simple test equipment. If one really becomes interested in this subject, it is highly recommended that a handbook on testing techniques be secured.

Sensitivity Tests

Sensitivity is a combined measurement having to do with both the noise quality and amplification ability of a receiver. The actual sensitivity value, usually expressed in microvolts, has no firm basis unless it is stated in conjunction with the receiver input impedance value, the receiver bandwidth, and an output signal-plus-noise to noise ratio (S plus N to N).

Basically, sensitivity is a measurement of this output S plus N sum divided by N ratio resulting from a given signal at the receiver antenna terminals. The receiver must be operating as a linear amplifying device and, therefore, AVC, noise limiter, and other circuits which modify receiver performance must be disabled during sensitivity measurements. The RF gain control must be set at maximum. The audio gain control setting can be varied to produce a reference reading as noted in Fig. 10-4 but cannot be changed once a measurement is started.

One of the main problems in sensitivity measurements is the correct termination of the generator to receiver antenna input and the receiver output to output meter interfaces. Several approaches are acceptable for coupling the signal generator to the receiver input terminals. Some measurements specify the use of the IEEE standard dummy load or antenna. Such a dummy load was supposed to represent a single-wire antenna between 150 kHz and 30 MHz which is very reactive at low frequencies and approaches a constant 400-ohm value at high frequencies. The use of such a standard has generally disappeared as receiver input circuits have been normalized at 50 or 70 ohms. Even so, the input impedance of a receiver is rarely very constant over a range of frequencies. Approximate sensitivity measurements can be accomplished by terminating a 50-ohm output signal generator directly in a nominal 50 or 70 ohm receiver input circuit.

Better results will be obtained if the generator and receiver are interfaced through a 6 db pad of the type shown in Fig. 10-5.

The pad insures that the SWR between receiver and generator will never exceed about 1.5:1. Thus, small variations in the receiver antenna circuit input impedance will not be of significance. The generator is assumed to have an output meter that reads the correct output voltage when it is terminated in the characteristic impedance of the generator

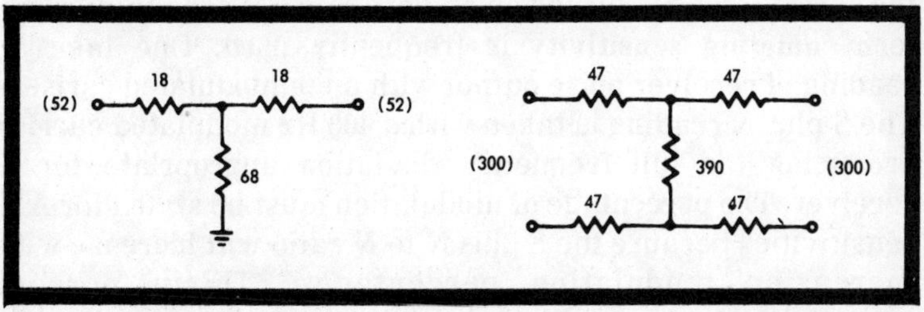

Fig. 10-5. 6 db pads are often used between generator and receiver when making sensitivity checks. Pads are shown for 52- and 300-ohm impedances.

(usually 50 ohms). With the use of a 6 db pad, the voltage present at the receiver terminals (the one that **should** be quoted for sensitivity values) is one half of the output voltage indicated on the generator output meter. The length of the cable connection between the generator and receiver should be kept as short as possible to avoid the introduction of capacitive loading. Generally, with cable lengths of 3 feet or less below 30 MHz, this effect will be insignificant.

Signal-Plus-Noise to Noise Ratios

The measurement of the S plus N to N ratio at the IF or AF output of the receiver appears simple but requires some clarification. Multimeters are calibrated for AC voltage readings based upon sine-wave values. Noise is not a sine-wave function. Therefore, there is some error when measuring S plus N to N values on either a simple multimeter or a vacuum tube or solid-state multimeter. For direct-driven multimeters, you must add about 1 db to the noise level indicated. For vacuum tube or solid-state multimeters which employ peak AC voltage rectifier circuits, you must subtract 3 db from the noise reading. Perhaps the best lesson that can be gained from realizing the error inherent in such measurements is that low S plus N to N ratios should be avoided. Much more accurate results will be obtained using any meter at 20 db ratios than at 6 db ratios.

Sensitivity measurements are necessarily tied to the modulation form being used, since the modulation generally affects the signal power input. For AM receivers, as noted in Fig. 10-5, the modulation, not the carrier, is turned on and off for sensitivity measurements. For CW or SSB measurements, the carrier alone is turned on and off. For FM sensitivities the term **quieting sensitivity** is frequently used. One takes a reading of receiver noise output with an unmodulated carrier. The S plus N reading is taken with a 400 Hz modulated carrier producing the full frequency deviation appropriate for a receiver. The percentage of modulation must be stated for AM sensitivities because the S plus N to N ratio will increase with increasing modulation percentages. Thirty percent modulation is generally taken as an industry norm to represent average speech condition.

The same general technique for selectivity measurement, Fig. 10-6, is applicable whether one is measuring the selectivity of a component, stage, or complete unit. A test signal of as great an amplitude as possible, but below the point of causing overload or distortion, is introduced at the center frequency (if known). A reference output level is obtained, and the output levels then are noted as the test signal frequency is varied either side of the center frequency. The frequency difference between points where the output level is the same determines the bandwidth at that **specific** relative output level. The output at the center frequency is generally taken as 0 db and the bandwidth determined at such points as half-power (3 db down, .7 of center voltage reading), 6 db down (.5 of center voltage reading), 60 db down, etc. The shape factor is the bandwidth at some attenuation point (60 db, for instance) divided by the bandwidth at one tenth the former attenuation point (6 db, for instance).

The closer the shape factor approaches 1, the steeper are the sides or "skirts" of a selective device. Shape factor has meaning only when the attenuation points are specified also. The limit to which attenuation can be measured depends upon the noise present in a system and the sensitivity of the output level meter used. For this reason, to accurately measure the signal selectivity of a receiver, which is determined by the IF selectivity, inject a signal directly into the IF strip and measure the IF voltage at the output of the IF strip using an RF probe and a voltmeter. It is feasible to measure the overall selectivity, injecting a signal at the receiver's antenna terminal and using an output meter connected to the audio output, for general maintenance purposes. When injecting a signal into the IF strip directly (at the mixer output circuit), the mixer stage and its associated oscillator preceding the IF strip should be made inoperable (remove the tube or remove supply voltage to the stage). The generator output should be coupled through a .001 mfd capacitor.

When measuring devices which have rather broad selectivity (6 kHz or more) are operating at radio frequencies, a modulated test signal can be used and the audio output level of a receiver can be measured. More accurate results, however, will be obtained if an unmodulated carrier is used and the voltage at the output of the IF strip or that across the

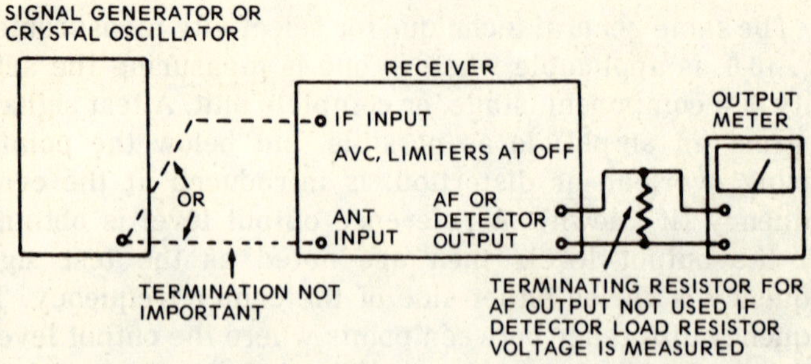

Procedure:
1. Tune the receiver to the generator frequency. Use a modulated generator signal and measure the AF output level, or with an unmodulated signal measure the detector load resistor voltage.
2. Use the highest generator level possible without causing overload. Use the RF-AF gain controls to set the reference output level.
3. Vary the generator frequency or tune the receiver in specific increments.
4. Plot the db output level decrease as a function of tuning increments.

Fig. 10-6. Procedure for making a bandwidth or selectivity measurement.

detector load resistor measured. Such a procedure is a necessity when testing very narrow filters with steep skirts such as SSB mechanical filters. When measuring narrow filters, the problem of adequate frequency calibration with simple instrumentation can become severe. It is sometimes better to vary the receiver frequency (when an overall receiver selectivity test is being made) rather than the test generator frequency, depending upon which instrument has the best calibration.

Sometimes, the calibrator oscillator in a receiver can be used as a test oscillator. Simply turn on the BFO (if the BFO frequency is centered on the IF passband) to obtain an audio output signal which then can be coupled to an output meter. Alternatively, the S-meter in most receivers can be used as the output indicator, if it functions with the AVC-AGC function switched off. Thus, one could measure selectivity without any external instrumentation. One drawback, however, is that the db calibration on S-meters is rarely accurate.

On the other hand, if you have a test generator with a well calibrated output, an alternate and more accurate procedure should be mentioned. You would obtain a reference output reading at the center frequency as mentioned before but use only a 1 to 10 volt test signal. The generator output would then be increased 3 db (1.4 times original voltage) or 6 db (2 times voltage) and its frequency (or that of a receiver) varied until the same reference output were again obtained. The procedure is repeated at whatever attenuation points the bandwidth is desired. In order to avoid the effects of dial backlash, the generator or receiver frequency is varied each side of the center frequency but always using the same dial rotation direction.

Audio circuit selectivity devices can be measured using a variable audio generator in the same manner as an RF signal generator. Also, it is possible to measure AF selectivity in a receiver by introducing a modulated test signal at the last IF transformer (after the main IF selectivity). Highly accurate results will not be possible if you are using an audio filter with an extremely narrow bandwidth, but general measurements are possible.

Frequency Calibration and Stability

Checks for receiver calibration and stability can be made with frequency counters on one extreme to just the receiver alone at the other extreme. The tests are simple with a counter. The receiver local oscillator frequency is observed for its stability over a period of time, and for calibration the counter reading of a test signal frequency is compared to that frequency indicated on the receiver dial. Using just the receiver alone, one can tune in a station of known good stability, such as WWV or CHU, on as many bands of a receiver as possible, zero-beat the incoming signal and observe how much the receiver tuning drifts as noted by the changing zero-beat point. Of course, the stability of the BFO oscillator is also involved but usually its stability will be far greater than that of the higher frequency variable local oscillator.

For dial calibration, one can place the BFO, if variable, at its set position and tune in a station of known frequency (WWV, CHU, crystal markers, calibrated heterodyne frequency meter output, etc.) to check the dial markings. With receivers having only crystal-controlled BFO oscillators for SSB reception, the BFO frequency is placed at a frequency on the steep portion of the IF selectivity curve. Zero-beat is often hard to distinguish, although it is usually possible to do so using a crystal calibrator with the receiving antenna disconnected. (The zero-beat must remain with either LSB or USB crystals switched in if the BFO is correctly aligned.) If a progressive frequency error is found across a band, or if a variable BFO must be set differently on different frequencies for the dial calibration to be correct, incorrect alignment of the receiver front-end and-or BFO may be indicated.

Within the extremes for the tests mentioned above, there are a number of other tests relating to calibration and stability that are worthwhile. Also, it is possible simply to refine some test procedures in order to make them a bit more accurate. First, however, there are various ways calibration and stability figures can be expressed. Calibration accuracy is usually expressed as a simple error in frequency (Hz or kHz) or as a percentage of the received frequency. Usually, the maximum error takes place at the highest received frequency.

Besides electrically-caused errors, there are almost always mechanical errors involved in the tuning mechanism. Stability can be expressed in frequency or percentage as a function of time (minutes, 24 hours, etc.) or as a function of some parameter change (10 percent line voltage change, for instance). Time is an important factor, since a receiver could have very poor short-time stability but excellent long-time stability. Rather than as a percentage figure, stability is often expressed in terms such as 1 part in 105. This would mean 10 Hz at 1 MHz, 100 Hz at 10 MHz, etc.

Many calibration and stability checks make use of aural zero-beating techniques. It must be appreciated that the restricted audio response of the AF stages at frequencies below about 100 Hz (and also that of the human ear) create a "gray" area of 100-200 Hz around zero-beat where you cannot be certain of exact zero-beat. One way to eliminate the human ear error to a great degree is to use an output meter or preferably an oscilloscope with a DC input amplifier to observe zero-beat. This does not make up, of course, for the restricted response in the receiver AF stages and it would be better still to connect the oscilloscope right at the detector output. There are a number of secondary-type stability checks which may be performed on a receiver, particularly inexpensive high-frequency receivers.

Stability is a function of input signal level. The test should be performed on the highest frequency usable with the AVC operative, RF gain at maximum, and AF gain set for a convenient output level. A signal generator test signal is zero-beat using the receiver main tuning. The signal generator output level is then varied from a barely audible signal to the maximum level the receiver can handle without overload. At each signal generator output level, the receiver is again zero-beat using the receiver main tuning. The excursion in the main tuning necessary to maintain zero-beat indicates the stability factor.

Another test is the gain control test, which can be done with either a strong received signal or a signal generator. The AVC is on, RF gain control at maximum, and the AF gain at a convenient level. A maximum amplitude (non-overloading) signal is fed to the receiver and the BFO set to produce some convenient output tone (800-1400 Hz roughly). The RF gain control is then rapidly turned down until the input signal is just audible. The change in the output tone frequency should be no more than just barely perceptible.

Warm-up drift and short-term stability are often of interest because equipment is frequently not left on continuously and poor short-term stability means a great deal of inconvenient retuning until the receiver finally stabilizes. To check warm-up drift, a well stabilized signal generator or crystal oscillator is used (not the receiver's calibration oscillator if built-in). The cold receiver is turned on and quickly zero-beat with the test signal. At specific intervals, perhaps 10 seconds, the receiver is again zero-beat and the excursion in main tuning setting necessary to maintain zero-beat is recorded versus time. Such a test is not technically a rigid one for many reasons (lack of line voltage stabilization, environmental control, etc.). Also, one does not isolate the circuit in the receiver causing the drift. Nonetheless, it is a good simple overall check.

Index